T0331645

Genes and the Agents of Life

The Individual in the Fragile Sciences
Biology

What are the agents of life? Central to our conception of the biological world is the idea that it contains various kinds of individuals, including genes, organisms, and species. How we conceive of these agents of life is central to our understanding of the relationship between life and mind, the place of hierarchical thinking in the biological sciences, and pluralistic views of biological agency.

Genes and the Agents of Life undertakes to rethink the place of the individual in the biological sciences, drawing parallels with the cognitive and social sciences. Genes, organisms, and species are all agents of life, but how are each of these conceptualized within genetics, developmental biology, evolutionary biology, and systematics? The book includes highly accessible discussions of genetic encoding, species and natural kinds, and pluralism about the levels of selection, drawing on work from across the biological sciences.

The book is a companion to the author's *Boundaries of the Mind*, also available from Cambridge University Press, in which the focus is the cognitive sciences.

This book will appeal to a broad range of professionals and students in philosophy, biology, and the history of science.

Robert A. Wilson was born in Broken Hill, Australia, and lives in Edmonton, Canada. He is the author or editor of five other books, including the award-winning *The MIT Encyclopedia of the Cognitive Sciences* (1999) and *Boundaries of the Mind* (Cambridge 2004).

Genes and the Agents of Life

The Individual in the Fragile Sciences
Biology

ROBERT A. WILSON

University of Alberta

CAMBRIDGE
UNIVERSITY PRESS

CAMBRIDGE
UNIVERSITY PRESS

32 Avenue of the Americas, New York NY 10013-2473, USA

Cambridge University Press is part of the University of Cambridge.

It furthers the University's mission by disseminating knowledge in the pursuit of education, learning and research at the highest international levels of excellence.

www.cambridge.org
Information on this title: www.cambridge.org/9780521544955

First published 2005

A catalogue record for this publication is available from the British Library

Library of Congress Cataloguing in Publication data

Wilson, Robert A., 1964–
Genes and the agents of life : the individual in the fragile sciences, biology /
Robert A. Wilson.
p. cm.
Includes bibliographical references and index.
ISBN 0-521-83646-8 (hb) – ISBN 0-521-54495-5 (pb)
1. Biology – Philosophy. 2. Genetics – Philosophy. 3. Developmental biology – Philosophy. 4. Evolution (Biology) – Philosophy. I. Title.
QH331.W558 2004
570'.1 – dc22 2004047297

ISBN 978-0-521-83646-3 Hardback
ISBN 978-0-521-54495-5 Paperback

For Miranda

Contents in Brief

Contents

List of Tables and Figures

TABLES

FIGURES

Acknowledgments

I began thinking about some of the material in this book in the mid-1990s as three emerging interests and activities slowly coalesced. Some of the metaphysical issues concerning the mind, particularly what I think of as the hard-nosed physicalist challenges to nonreductionist views posed by Jaegwon Kim and David Lewis, had been passed by too quickly in my first book, *Cartesian Psychology and Physical Minds*. I had also started developing, slowly and through trial-and-error teaching, some background in biology and the philosophy of biology. And, finally, I was forced to think about the diversity of views within cognitive science through my role as general editor for *The MIT Encyclopedia of the Cognitive Sciences*. It took a few years to find a thread linking these interests and issues, and for the project of which this book is a part, *The Individual in the Fragile Sciences*, to be articulated.

The project itself was initiated while I held a fellowship at the Center for Advanced Study at the University of Illinois, Urbana-Champaign, in the spring of 1998. The Center also played a broader supportive role for me throughout my time at Illinois, for which I am grateful. While I wrote very little material that has found its way into this book during that period, I broadened my background in biology with the assistance of a fellowship in the Program for Liberal Arts and Sciences Study in a Second Discipline at Illinois. I thank Jay Mittenthal for serving as my advisor for this fellowship, and the faculty who put up with my snooping around their courses in cell biology and molecular evolution. The Cognitive Science Group at the Beckman Institute at Illinois provided a stimulating intellectual home that regularly transgressed disciplinary boundaries of all kinds. I thank my colleagues there and in the Department of

Philosophy – especially Gary Dell, Gary Ebbs, Steve Levinson, Patrick Maher, Greg Murphy, and Fred Schmitt – for fostering a constructive and welcoming academic environment in which a somewhat open-ended project could be undertaken.

Most of the writing for the project has been done while I have been at the University of Alberta these past two or three years, and with generous release time provided by grant 410-2001-0061 from the Social Sciences and Humanities Research Council of Canada. As the project developed, it became clear that it could not be completed within a single book, and much of the last year has been spent making both this book and its companion, *Boundaries of the Mind* (Cambridge, 2004), walk the line that independent but related books must. I thank my departmental chair, Bernie Linsky, for his flexibility in assigning my teaching load, giving me the whole of 2002 to concentrate on both books. My thanks also to my research assistants over the last two years – Ken Bond, Jennie Greenwood, Li Li, and Patrick McGivern – and to members of an upper-division and graduate course in the spring of 2003 who grappled with, and usefully critiqued, material from both books.

For feedback and encouragement along the way, I would like to thank Jim Brown, David Castle, Peter Godfrey-Smith, Ben Kerr, Alex Rueger, Kim Sterelny, and David Sloan Wilson. Audiences at the University of Alberta and the University of Calgary have provided me with useful feedback on the material in Parts Two and Four, and I trust that they are better for it. I owe a special debt to Michael Wade, whose detailed comments on the penultimate draft gave me reason to make the final revisions more substantial than I had hoped they would be, not least of all because he pointed me to large swaths of work of which I would otherwise have likely remained blissfully ignorant. Likewise, several of the symposia at the 2003 meetings of the International Society for the History, Philosophy, and Social Studies of Biology in Vienna – particularly those on rethinking the history of essentialism and on the extended phenotype – gave me more food for thought in making final revisions. Once again, the editorial team at Cambridge – Stephanie Achard, Jennifer Carey, Shari Chappell, and Carolyn Sherayko – have been a pleasure to work with throughout and have improved the final manuscript considerably. I would also like to thank four reviewers for Cambridge University Press who provided general feedback and guidance on the overall shape of the project, and Terence Moore at Cambridge for his editorial leadership.

Some of the chapters draw on and develop material that I have published elsewhere. I would like to acknowledge and thank the publishers for permission to include material drawn from the following sources:

Chapter 5: "Realism, Essence, and Kind: Resuscitating Species Essentialism?," in Robert A. Wilson (editor), *Species: New Interdisciplinary Essays* (Cambridge, MA: MIT Press, 1999), 187–207.

Chapter 6: "The Individual in Biology and Psychology," in Valerie Gray Hardcastle (editor), *Where Biology Meets Psychology: Philosophical Essays* (Cambridge, MA: MIT Press, 1999), 357–374.

Chapter 7: "Some Problems for 'Alternative Individualism,' " *Philosophy of Science* 67 (2000), 671–679. © 2000 by the Philosophy of Science Association. All rights reserved. The University of Chicago Press.

Chapter 9: "Test Cases, Resolvability, and Group Selection: A Critical Examination of the Myxoma Case," *Philosophy of Science* 71 (July 2004), 380–401. © 2004 by the Philosophy of Science Association. All rights reserved. The University of Chicago Press.

Chapter 10: "Pluralism, Entwinement, and the Levels of Selection," *Philosophy of Science* 70 (July 2003), 531–552. © 2003 by the Philosophy of Science Association. All rights reserved. The University of Chicago Press.

Edmonton, Alberta
October 2003

PART ONE

INDIVIDUALS, AGENCY, AND BIOLOGY

1

Individuals and Biology

1. INDIVIDUALS AND THE LIVING WORLD

What are the agents of life? This book is a partial answer to this simple-sounding, yet puzzling, question. In this first chapter, I shall unpack what is built into this question and introduce some of the issues that answering it will lead us to explore.

The living world not only surrounds us physically, but its denizens also occupy much of the content of human thought and action. These agents of life range from the plants and animals that fill our homes and domestic lives, to those we consume as part of our ecological regime, to organisms of all types: blue whales, dolphins, chimpanzees, dogs, fungi, flowering plants, rainforests, bacteria, viruses, sponges, tapeworms, and so on.

The living world and the agents of life that constitute it excite the full range of our passions – love, wonder, joy, fear, and disgust. Our interactions with them have inspired human artistic expression from the earliest cave drawings to late-twentieth century experiments in bio-art.

Part of what impresses us, what leaves a mental mark, is the fact that we are not simply immersed in the living world but part of it. We are each subject to its vicissitudes, such as disease and death, and each of us owes our own existence to the activities of members of the living world most like us: other human beings. Human beings are the agents of life that preoccupy most of us, most of the time, and it is more than naive anthropomorphism that places us and our own peculiar qualities at the heart of many of our representations of the living world.

But there is more to what catches our eye about the living world. It is rich, complex, and diverse. Both its particularity and its lawfulness are

3

the basis for at least a minor kind of awe, one that can stem from our everyday interactions with the rest of the living world. If you have ever carefully observed large mammals in their natural environments, played with a young kitten or puppy, watched a flower bloom or a bee work its way across the stamens in the flower, or learned how rapidly viruses and bacteria can multiply and change their structure, then you might have experienced the little "Wow" that I am trying to evoke.

There is a simplicity to our common-sense thinking about the richness, diversity, and complexity to the living world that is taken for granted in both common sense and the biological sciences. The living world is made up of living things, and living things are agents. We think about the dimensions of the living world, and the vague and contentious boundaries to it, in terms of the individual agents that the living world contains. These individuals range in size from single-celled organisms that can only be viewed through moderately powerful microscopes to giant sequoias that reach hundreds of meters into the sky and whose full physical scale, both above and below the ground, cannot be observed in any single, direct way. The living world includes individuals that vary massively in their longevity, with lifespans of minutes to those of hundreds of years; in their general strategies for living (for example, plants versus animals); in their internal complexity; and in their relationship to one another and their environments.

In short, when we think of the living world we think of the individual agents in it, the properties those individuals have, and the relationships they enter into, both with other living things and with the nonliving world. We think of life as we think of the mind – as tied to and delimited by agents. Neither life nor mind float free of agents, but are, in some sense, features that individuals either have or lack.

There is a flip side to this tie between life and agency. Life, like mind, does not simply belong to agents, but is more intimately woven into their fabric. Life and mind determine what it is to be an agent. Life and mind are, in some sense, *inside* individuals. They are deep features that make an important difference to the kind of individuals that have them, and they cannot be removed from any individual without changing their status as agents.

2. LIFE AND MIND

This raises the question of just how closely life and mind are related. Ancient and modern theories of both life and mind imply an intimate relationship between the two, an intimacy reflected in the etymology of classical and modern languages. For example, Aristotle distinguished between three kinds of soul that living things can possess – vegetative, animal, and rational. Herbert Spencer, the great integrator of nineteenth-century philosophy and science, viewed life and mind as being of a piece, and took a treatment of one to be incomplete without a treatment of the other. The words usually translated as "soul" from Sanskrit, Greek, and Latin – *atman, psyche,* and *anima* – all have the connotation of breath, something that fills a living thing and is necessary for its survival. The contemporary physicist and popular science writer Fritjof Capra says "[d]escribing cognition as the breath of life seems to be a perfect metaphor," and identifies life and mind.[1]

Today these views of life and mind are likely to be seen as antiquated or quaint, and met with corresponding bemusement or the impatience that leads to contempt. We treat life and mind as independent features that an individual either has or lacks. The study of life and mind has been compartmentalized into, respectively, the biological and the cognitive sciences. This sort of disciplining of the domains of life and mind, however, was contingent rather than inevitable. While it has created the opportunity for deep insights into both life and mind, it has also produced its own blind spots. One of these concerns the role and conception of agents in thinking about life and mind. Let us begin comparatively and consider two ways in which these roles differ.

First, in the case of the mind, we have a clear paradigm for the sort of agent that minds belong to. They belong to rational beings, and, for better or worse, we view human agents as rational beings. This is not to deny that we often think of other kinds of agents as having minds. For example, intelligent robots or computers, such as HAL in *2001: A Space Odyssey,* God (on at least most conceptions), and nonhuman animals all have minds of some kind. But human beings are the gold standard in that the minds of these other agents are typically conceptualized as being somewhat like those of human agents but diminished or enhanced in some or all of their characteristics. The focal role that human rational agency plays in our common-sense thinking about minds survives in the sciences of the mind – from artificial and computational intelligence to comparative psychology – where minds are conceptualized in terms of categories, such as perception, learning, decision, and memory. Although these straddle the divide between human and nonhuman cognition, again they have their paradigm existence in human agents.

In the case of life, we also have a paradigm conception of what sort of agent has it, one that encompasses human agency but is somewhat broader. Our paradigm examples of living things are organisms. We are organisms, true enough, but it is also true that we are simply one kind, perhaps one very special kind, of organism. Our place in the living world is not as central as it is in the domain of cognition, a point reflected in the diminished role that human agency has within the biological sciences relative to that in the cognitive sciences.

A second difference between the role and conception of agents in thinking about life and mind concerns the physicality of individuals. By this I mean their boundedness in space and time, their material composition, and the role of dimensions of physical continuity in the survival of individual entities. Living agents have a high level of physicality. They are born (come into existence) and die (pass out of existence), they have a particular material structure important to what they can and cannot do, and their identity over time as the very same living thing – and not just the same kind of living thing – depends heavily on their physical continuity during that time. Cognitive agents, by contrast, have a low level of physicality. They are often thought of as having an essence that can be separated from their physical embodiment. This is so, not only in religious thought that embraces the survival of the soul after the death of the body, but also in the familiar fantasies of science fiction in which minds can be stored as information and beamed from one physical medium to another. The minimal physicality of cognitive agents is also manifest in the traditional view within artificial intelligence, which takes physical embodiment to be an add-on to a cognitive agent, something into which the artificial intelligence is injected after an agent has been created or established. It is part of what philosophers often think of as a Cartesian tradition that sees minds and bodies as operating in two different worlds, the mental and the physical.[2]

3. AGENTS: BIOLOGICAL, LIVING, AND OTHER

So far I have moved freely between talking of organisms as paradigms of living things, as individuals in the biological sciences, and as agents. It is time now to sharpen our focus and introduce some terminology. The central notions here will be those of a *living agent* and a *biological agent*.

I intend to characterize an agent in quite a general way: an agent is an individual entity that is a locus of causation or action. It is a source of differential action, a thing from which and through which causes operate. Consider some concrete examples of agents that are physical, biological, and social in nature.

Agents in the physical world, or physical agents, include very small things, such as elementary particles and atomic elements; ordinary physical objects of the sort that you can see with the naked eye or manipulate with your body, such as balls, tables, and rocks; and larger and more distant objects, such as tectonic plates and stars. Agents in the biological world, or biological agents, include proteins, genes, cells, organisms, demes, species, and clades. Social agents include individual people, but also groups of people, institutions, networks, and larger systems that consist of these other agents organized in particular ways.

The notion of an agent is linked, but not identical, to that of a cause. Agents are individuals, and causes often are not. I would be content to contrast individuals with other kinds of things that we might invoke as "the cause" in a given instance, such as forces and fields, processes and events, and properties and states. But I do not want to be legislative about this, and there are certainly ways in which we can and do think about, for example, certain winds or a particular magnetic field as an individual. Giving them a proper name, such as the North Wind or Hurricane Eliza, or personifying them more generally, are two ways of treating forces and fields, for example, as individual agents.

Crucial to being an agent, in the broad sense I intend, is having a boundary, such that there are things that fall on either side of that boundary. This notion of an agent should both make it clear why organisms are paradigmatic agents and why bodily *systems*, such as the digestive system, and biochemical *pathways*, such as the integrin signaling pathway that mediates cell adhesion, might be considered biological agents. As these examples suggest, agents sometimes operate as biological mechanisms: they have functions to perform in the context of some larger agent, and in turn contain further agents (such as the stomach and cadherins, in the two above examples) that perform contributory functions.

I find it compelling to think of these agent-marking boundaries as spatial and temporal, and so view agents as having both spatial and temporal beginnings and endings, as well as spatio-temporal continuity throughout their existence. Yet some of the agents most commonly invoked across cultures in explanations and accounts of our personal experiences are thought to be nonmaterial: from God, to angels, to ghosts, to ancestral spirits. Any view of agency needs to say something about such putatively nonmaterial agents, but we need not do so in this introductory chapter.

Consider then (and for now) just physical agents, agents that have spatial and temporal boundaries, a material composition, and continued existence in space and time. Biological agents are one kind of physical agent. What of *living* agents? Put simply, these are biological agents that are living. Living things are often characterized in terms of one or more of the following properties: they have a metabolism, grow, contain adaptations, evolve, or have heterogeneous, specialized parts. The relationship between the concepts of organism, agency, and life is the topic for several chapters in Part Two. For now, I simply appeal to these properties as a way to provide a fix on what a living agent is, and note that, with the exception of organisms, we might reasonably question whether *any* of the examples of biological agents that I provided previously – proteins, genes, cells, demes, species, and clades – are living agents in and of themselves.

There are two important features of the way in which we think about biological agents, including living agents, particularly in the contemporary biological sciences.

First, biological agents are often conceived as forming a hierarchy of increasingly inclusive entities, starting with very small biological agents and ending with larger entities comprised of the agents with which we began. As the philosopher Todd Grantham says,

Life on earth is hierarchically organized. The biotic world consists of many 'levels' with the entities at each higher level composed of lower-level entities. Groups of cells form the tissues and organs out of which organisms are constructed, and organisms form various kinds of groups such as kin groups, populations, and species.[3]

This hierarchical thinking is ubiquitous in the biological sciences, and it extends to include not just agents (in my sense) but processes, events, properties, and states. In all of these cases, our default view is a sort of realism about these hierarchies. They and the individuals they contain are a part of the fabric of the world, rather than simply a product of our ways of thinking about the world, something we discover rather than invent.

Second, it is common to distinguish kinds of biological agents from one another. For example, some are physiological (cells), others are genetic (segments of DNA), some are ecological (predators), others evolutionary (species). These specific kinds of biological agents are also thought to be organized hierarchically. Together with the fact that the resulting more specific lists of biological agents are almost always distinct, this suggests that there is no single listing of "the" biological agents there are in the world.

TABLE 1.1. *The genealogical hierarchy and two versions of the ecological hierarchy*

Genealogical	Ecological (1)	Ecological (2)
Monophyletic taxa	Regional biotas	Biosphere
Species	Communities	Ecosystems
Demes	Populations	Populations
Organisms	Organisms	Organisms
Chromosomes	Cells	Cells
Genes	Molecules	Molecules

Source: Redrawn from Table 6.3 of Niles Eldredge, *Unfinished Synthesis* (New York: Oxford University Press, 1985).

Consider a version of the distinction between genealogical and ecological hierarchies initially introduced by the paleontologists Niles Eldredge and Stanley Salthe (Table 1.1).[4]

The genealogical hierarchy contains entities that form historical lineages and give rise to patterns of ancestry and descent. The ecological hierarchy, by contrast, orders entities that play some sort of economic or functional role in the activities of life. As the distinction between two possible ecological hierarchies suggests, there are different ways to individuate the entities in these hierarchies, here turning on whether, as in the middle column, we restrict our ecological hierarchy to living things, or whether we take it to also include the abiotic environment, as in the right-most column.

Both the role and nature of hierarchical thinking within the biological sciences, and the idea of pluralism about biological agents, are topics that will occupy us further throughout *Genes and the Agents of Life*. As a way of illustrating how both topics are engaged by alternative conceptions of the individual in the biological sciences, I turn next to consider the long shadow cast by the Aristotelian view of the natural world and challenges to it in relatively recent biological thinking.

4. SPECIES AND NATURAL KINDS: THE ARISTOTELIAN SHADOW

I have already noted the obviousness of individual organisms when one looks at or reflects on the organization of the biological world. Only slightly less perspicuous a feature of that organization is that organisms are not randomly assorted throughout nature but cluster in groups whose members are similar to one another, or are of a kind. Plus or minus a bit, biological species – whether they are human beings, domestic dogs or cats, or robins, to take four of the species most often invoked in common-sense thought and talk – strike us as a form of organization in nature, a natural kind. As recent work in folk biological taxonomy suggests, "[h]umans everywhere classify animals and plants into specieslike groupings that are as obvious to a modern scientist as to a Maya Indian."[5]

Given the naturalness, at least to us, of this level of organization in the biological world, it should be no surprise to learn that the idea that individual organisms belong to natural kinds, species, has a long history. It is often associated with Aristotle, and some conception of biological species has remained central to the history of Western thought about the structure of the biological world since his time. The general metaphysical categories of individual, species, and genus, and the relationships between them, and between them and the rest of reality, play a central role within Aristotle's metaphysics. Many of the examples that Aristotle uses to illustrate these general categories are biological in nature. Since the general outlines of Aristotle's views have remained influential throughout a range of other, sometimes quite radical, changes in metaphysical views, it will pay to have at least that outline before us in thinking about individuals and species in the biological world.

Individual organisms are paradigmatic instances of what Aristotle calls *substances*, the true subjects in the world, the things of which everything else is predicated but which are not themselves predicated of other things. Substances are the focus of the study of metaphysics, which strives to understand their nature or essence. Throughout his writings, Aristotle recognizes some of the similarities between individuals and what he calls species of individuals. In his *Categories* he goes so far as to distinguish explicitly between *primary* substances, individuals, and *secondary* substances, of which species, including biological species, are paradigmatic instances. In so doing he underscores the importance of these similarities. An individual human is an example of primary substance, and to predicate "human" of that person is, in part, to define what sort of thing that individual is, in a way that predicating color or height of him or her does not. To say that humans are animals, that is, to predicate the genus of the species, is to do just the same thing. Thus, for Aristotle, both species and genus are secondary substances, with species being "more truly substance than the genus."[6]

Aristotle's metaphysical picture implies that the biological world is hierarchically structured, and that this structure constitutes a way in which the biological world is unified. There is a single way in which it hangs together, represented by the taxonomic schemata of evolutionary biology. On the Aristotelian view, species are a fixed part of the order of things. Although the fixity of species was one of the central ideas challenged by Darwin's *On the Origin of Species*, a modified essentialism about species, one that viewed them as natural kinds, albeit with essences that could change over time, has largely been taken for granted throughout the history of biology and philosophy.

Biologists and philosophers of biology have, over the past thirty years, challenged this Aristotelian framework, particularly its essentialism about species and its unificationism about the order in the biological world. These challenges and the resulting alternatives to essentialism and unificationism – namely, the idea that species are individuals and species pluralism – have been so successful that they have usurped the traditional view of species in contemporary philosophy of biology. Species are not simply comprised of individuals but are themselves individuals, not natural kinds. And there is not any one order of things in the biological world, represented by "the" species concept and its place in the Linnaean hierarchy, but many such orders, represented by various species concepts.[7]

Both the thesis that species are themselves biological individuals, and the claim that we should be pluralists about species concepts, deserve more articulation than that provided by my bare summary. But it should be clear already that these views are integral to a variety of issues about the nature of the biological world and our thinking about it, and that they have been viewed as such by their proponents. For example, if species are themselves individuals, rather than natural kinds, then individual organisms are parts of, rather than members of, species, and essentialism about species membership looks something like what philosophers call a *category mistake*. And if pluralism is true, then attempts to articulate "the" species concept can never succeed, for there is no single biological reality for such a concept to map to.

5. PLURALISM, REALISM, AND SCIENCE

Pluralism has considerable vogue within contemporary philosophy of biology and biology itself. I have already mentioned pluralism about species concepts, a pluralism that can be readily extended to the various more general approaches within systematics (for example, cladistics, phenetics). But one can find pluralistic views in many other areas of biology: in debates over the levels of selection, in disputes about the concept of the gene and the role of genes in evolution and development, and in controversies concerning adaptationism as a research strategy in evolutionary biology. For many of the topics that I shall discuss in the remainder of *Genes and the Agents of Life*, including the four just mentioned, pluralism is not simply a possible position but one defended by many of the leading figures in the field.

There are diverse motivations for this plurality of pluralist views, several of which transcend the particular issues to which pluralism is a response. For philosophers, pluralistic views often mark a departure from traditionally dominant views within the philosophy of science. These include the view of science typically attributed to the logical positivists, according to which the sciences are unified by the hierarchical relationship that holds between them, with the world that the sciences describe featuring entities that stand in a parallel hierarchy, from the very small to medium-sized dry goods to the truly large. And they include a tradition originating in the seventeenth-century mechanical philosophy, what we might call the Cartesian-Newtonian view of the universe as a gigantic machine. On both views, the unity of science is derived from the ways in which small things fit together to form larger things.

Such views are seen, I think rightly, as imposing a sort of straightjacket on the biological sciences, forcing their conformity with the physical sciences taken as a paradigm within the philosophy of science until the last thirty years. In introducing a recent collection of his essays in the philosophy of biology in which pluralism is a recurring theme, Philip Kitcher characterizes the early 1970s as a time when "philosophy of science clearly meant philosophy of physical science," a characterization that echoes that of other leading philosophers of biology, such as David Hull and Elliott Sober. Pluralistic views of explanation, as well as theories, taxonomies, and methodology in science have appealed to those reflecting on the nature of biology in part because they make this a time of the past, allowing the biological sciences to be assessed on their own terms, rather than in the image of physics.[8]

So one motivation for pluralism within the philosophy of biology might be characterized, in the most literal sense, as reactionary in rebelling against dominant traditions within the philosophy of science. But pluralism carries with it a more positive view of the nature of biological reality, of the biological world as more complicated, various, and messy than even our sophisticated views of theories, explanations, and kinds have allowed. Pluralism aims to more adequately capture this complexity and the corresponding diversity within explanatory practice in the biological sciences. Writing on pluralism about species, John Dupré says that "the more we have learned about the complexity of biological diversity, the clearer it has become that any one theoretically motivated criterion for taxonomic distinctness will lead to taxonomic decisions very far removed from the desiderata for a general reference scheme." Likewise, Kitcher's pluralism about the concept of function in biology aims to do justice to how that concept is used in a broader range of biological sciences than just evolutionary biology.[9]

Both Dupré and Kitcher see their pluralistic views as not only doing more justice to the diversity and complexity of the biological world, but also as providing some recognition of the *social* dimension to scientific practice and theory. It is not simply that the sciences have a social history that influences the conceptual tools that they employ. The practice of science is also subject both to the social division of labor and to regulation by particular social values. Pluralistic views are seen as the natural result of a philosophy of science sensitive not only to what has been called the postpositivist naturalistic turn but also to the interdisciplinary trading zone between the philosophy, history, and social studies of science. Thus, both Dupré and Kitcher have devoted much of their research effort to ways in which the biological sciences are intertwined with social issues and agendas: Dupré to the political pitfalls of reductionism and evolutionary psychology, and Kitcher to the Human Genome Project and eugenics.[10]

Scientists fight different battles. One of the appeals of pluralism for biologists themselves has been that it provides a diagnosis for resolving or perhaps altogether avoiding a debate at an apparent impasse. Consider one of these, the debate over the levels at which natural selection acts. In advocating a form of pluralism about the levels of selection, the entomologists Andrew Bourke and Nigel Franks say that "colony-level, group, individual, and kin selection are all aspects of gene selection. This means that the practice of attributing traits to, say, either colony-level selection or kin selection is illogical." Here there is the feeling that participants in such debates are talking past each other, or that their disagreements are merely semantic. In this context, the adoption of pluralism is a way to represent not so much diversity but underlying, core agreement within the biological sciences. It constitutes the diagnosis of a hidden consensus on which the science can build, bypassing what might otherwise be taken to be irresolvable disagreement limiting scientific progress.[11]

For both philosophers and scientists, pluralism has been defended together with at least a tempered form of realism about science. Realists hold, roughly, that the sciences not only aim at, but at least sometimes achieve, accuracy in how they depict the world. This is so whether we consider "observational" aspects of science, such as whether an organism has blue, brown, or white eyes or the particular temperature that a liquid has at a given time, or its "theoretical" claims that transcend what can be observed. When Mendel posited particulate "factors" inside his pea plants that were causally responsible for the patterns of character traits that they produced, he was making a theoretical claim in this sense, one that realists view as capturing part of an unobserved structure to reality, that part which we now refer to with the concept "gene" and associated concepts. As such, realism contrasts both with empiricism, which ascribes more significance to variations of the divide between the observational and the theoretical, and social constructivism, which emphasizes historical and social aspects of science over its putative search for Truth.

While pluralism takes some steps from traditional realism toward social constructivism, its proponents are clear to distinguish their view from more radical, relativistic forms of constructivism. Like other middle-ground positions, pluralistic realism faces pressures from views on either side of it: from traditional realism (against its pluralism) and from constructivism (against its realism). This issue will loom large as we turn to particular pluralistic views: about organisms and species (Part Two), genes and developmental resources (Part Three), and the levels of selection (Part Four).

6. WITHIN THE EVOLUTIONARY HIERARCHY: GENES, ORGANISMS, GROUPS

The "Modern Synthesis" refers to the amalgamation of distinct biological disciplines that emerged during the 1930s and 1940s. At the heart of the Synthesis was the putative integration of two traditions within biology – evolutionary theory, stemming from Charles Darwin, and genetics, originating with Gregor Mendel. One of the architects of the Synthesis, the geneticist Theodosius Dobzhansky, has said that nothing in biology makes sense except in the light of evolution. While this no doubt exaggerates the significance of the fact of evolution and the place of evolutionary theory within the practice of the biological sciences, it serves as a reminder of the centrality of evolution and evolutionary theory to a full understanding of the biological world, something fundamental to the Synthesis. Our paradigmatic biological agents, organisms, form lineages, bear adaptations, and are differentially selected accordingly to their level of fitness. These are all features of organisms that make sense in the light of evolution. Organisms are evolutionary agents, and this fact about them underlies many explanations of the facts that we can observe about them.[12]

But as the genealogical hierarchy in Table 1.1 makes clear, organisms are not the only evolutionary agents. This raises the question of the place of organisms in the hierarchy of evolutionary agents. This question is typically addressed as a central part of the issue of the level or levels at which natural selection operates. A brief discussion of it here will illustrate that how one conceives of biological agents, and the role that one ascribes to those individuals, structures and constrains theory and practice within the biological sciences.

In the traditional Darwinian theory of natural selection, the individual organism plays the central role as the agent on which natural selection operates. Organisms are the individuals that bear phenotypic traits, that vary in their fitness within a population, and that, as a result, are selected for over evolutionary time. Organisms are the bearers of adaptations, such as thick coats in cold climates, or porous leaves in humid climates. They are the units of selection, the level at which selection operates. On Darwin's own view, units larger than the individual, such as the group, were for the most part unnecessary, and units smaller than the individual, such as the gene, unknown.

By contrast, in the postsynthetic view of evolution by natural selection often glossed in terms of the concept of the selfish gene, individual organisms play a very different role. On this view, genes rather than organisms are the agents of selection. They come to play many of the roles, and have many of the features, ascribed to organisms on the traditional Darwinian view. On this view, organisms are not much more than ways in which genes get to propagate themselves. In terms that Richard Dawkins uses, they are the *vehicles* in which the real agents of selection, genes, the *replicators* in the story of life, are lodged. As Dawkins says in *The Selfish Gene*, "A monkey is a machine that preserves genes up trees, a fish is a machine that preserves genes in the water." Genes are the ultimate bearers of adaptations, coding for the phenotypes that are expressed in the organisms they build. Variations in fitness between genes provide the basis for the process of natural selection. Furthermore, not only is the individual organism no longer the agent of selection, but as Dawkins has also argued, it is only an arbitrary boundary for phenotypes. Phenotypes are extended, reaching into the world beyond the organism, rather than being organism bound.[13]

Each of these conceptions of the role of the organism in the theory of natural selection carries with it implications for a number of issues to which that theory is central. I shall mention just three here: the problem of altruism, higher-level selection, and the integration of ecology and genetics.

Altruistic phenotypes and behaviors are those that decrease the relative fitness of the organisms that bear them. On the traditional Darwinian view of natural selection, focused as it is on organisms and their reproductive success, the existence of altruism represents a puzzle. If organisms are the agents on which natural selection operates, then natural selection will select those organisms that have a relatively higher level of fitness. Thus, organisms adapted to increase the relative fitness of other organisms, that is, decrease their own relative fitness, could not evolve through this process, except as incidental by-products. The problem of altruism, on this view, is how to explain the existence of organism-level altruism.

On the gene-agency view of natural selection, by contrast, this problem does not exist. Or, rather, it is solved by showing how altruism is a result of the process of natural selection operating on genes and maximizing their reproductive success. For copies of the same genes can exist in different organisms, and so at least some forms of altruistic behavior would be predicted by a gene-agency view of natural selection where organisms share significant proportions of their genes, such as when they are kin.[14]

A second issue for which the individual- and gene-agency views of natural selection have implications is how important higher-level selection is in shaping the tree of life. Such selection operates on agents larger than the individual, anything from temporary dyads, to demes, to species, to whole clades. Both the traditional Darwinian and the gene-agency view are circumspect about the need to posit higher-level selection and do so only when explanation at their preferred level is not empirically adequate. On extensions of the traditional Darwinian view, higher-level agents such as species or clades are conceptualized very much as organisms in that they are seen as sharing many properties ordinarily ascribed to organisms. On the gene-agency view, groups, species, and clades are simply larger pools of genes, different sized vehicles if you like, but they are never truly the agents of selection.[15]

Finally here, consider two recent ways in which biologists have conceptualized the further integration of ecology and genetics in light of the idea that phenotypes extend beyond the boundary of the individual organism. On the first, developed by Kevin Laland, John Odling-Smee, and Marcus Feldman, the key notion is that of *niche construction*, with ecological niches being both developed by organisms and passed on across generations along with genetic material. This involves generalizing the notion of inheritance to encompass environmental resources, and recognizing the role that ontogenetically constructed environments themselves have in influencing gene flow over evolutionary time. An alternative way forward, developing the notion of *community genetics*, originally postulated by the biologists James Collins and Janis Antonovics, focuses on the mutual influences of genetic variation and ecological interactions in multispecies populations. Community genetics attempts to model what have been called "indirect genetic effects," the effects that genes in one individual have on the genes and phenotypes of other individuals, and the relationships between population structures and the dynamics of gene flow more generally. Some view community genetics as providing support for a conception of communities as superorganisms. Both of these views follow Dawkins in moving beyond the boundary of the organism in conceptualizing where phenotypes begin and end, yet neither implies that selfish genes are the ultimate evolutionary agents.[16]

I shall undertake a more extended discussion of evolutionary hierarchies in Part Four, focusing particularly on the roles that groups play in the process of evolution by natural selection. My point here is that just where and how we see evolutionary agency gain purchase in our biological hierarchy – via individuals or via genes – has ramifications both for how standard problems are articulated and for what view we adopt of other forms of evolutionary agency.

7. BIOLOGY AND THE FRAGILE SCIENCES

The concept of the individual is central to how we think about the mind, living things, and the social world. The sciences that concern each of these domains – the cognitive, biological, and social sciences – have developed independently. One of my working assumptions, one that has surfaced in the first half of this chapter, is that we would often do well in thinking about individuals to conceptualize each of these sciences (better: cluster of sciences) as part of a broader range of sciences, the *fragile sciences*. In this section, I want to say something about biology within the fragile sciences as well as something about the neologism "the fragile sciences" itself.

Psychology and the social sciences are sometimes discussed together under the rubric of the *human sciences* or the *behavioral sciences*. But these are terms more likely to be used by those studying rather than practicing the sciences such terms aim to encompass. It has been even more unusual to further extend such conceptions to encompass the full range of biological sciences as well. The divide between the cognitive and social sciences, on the one hand, and the biological sciences on the other, is institutionalized in the divisions between university faculties, funding agencies, publishing venues, and professional organizations. This separation does not simply exemplify institutional demarcation but reflects theoretical asymmetries that are thought to exist between these sciences, asymmetries that serve as the basis for thinking of the biological sciences as foundational for, and more basic than, the cognitive and social sciences. In each case, biology stands to psychology and the social sciences very much as physics stands to it, a point I shall illustrate where appropriate.

The first sense in which the biological sciences have been thought of as more fundamental than their cognitive and social counterparts is their more general scope. They offer theories that apply not only to agents with minds (the cognitive sciences) and to agents that develop societies (the social sciences) – human agents in the first instance, and nonhuman animals derivatively – but to living things more generally. Physics, in turn, is more general in scope than is biology. These scope differences are one basis for thinking of psychology and the social sciences as "special sciences": the individuals they treat are special kinds of individuals that belong to a larger class about which there can be systematic generalizations.

A second source of asymmetric demarcation is that the biological sciences were disciplined earlier than were psychology and the social sciences, in the eighteenth century and the first half of the nineteenth century, rather than in the century that followed. That is, the biological sciences came to be conceptualized as having a distinctive subject matter prior to that of the psychological and social sciences, such that the latter came to be both justified and formed against preexisting biological sciences. Medicine and anatomy, and natural history and plant and animal taxonomy, were all well-established disciplines in much of the eighteenth century, one hundred years prior to the early murmurings of psychology and the social sciences in psychophysics, experimental psychology, and sociology. Developments in the biological sciences that coincided with the formation of psychology and the social sciences as distinct disciplines close to the turn of the twentieth century, such as cell biology, genetics, and developmental biology, occurred within an institutional safety net for biology as a field of inquiry.

A third way in which biology has been set apart from psychology and the social sciences is its perceived closeness to, and continuity with, the physical sciences, paradigmatically chemistry and physics. This perception has been promoted within the physical sciences themselves through reductionist views of core biological notions, such as that of the gene as a sequence of DNA, and the development of physics-based approaches to biology, such as general systems theory, information-theoretic cybernetics and, more recently, self-organizing systems theory. This proximity between biology and physics is reflected in disciplines that bridge between the two, such as biochemistry, bioinformatics, and computational biology. And methodological continuities and similarities, rather than disjunctures and differences, have been emphasized, ranging from experimental, mathematical, and analytical techniques to the search for general laws and constraints.

The significance of these differences – in scope, in temporal priority, or in proximity to the physical sciences – between the biological and the other fragile sciences, particularly whether they provide reason for treating them distinctly, remain open issues. Uncertainty here derives in part from the sheer diversity within each of the biological, cognitive, and social sciences. For example, large tracts of biology, including much of evolutionary ecology, optimality modeling more generally, and functional morphology, appeal to teleology, treating organisms as if they were artifacts designed to achieve specific goals in specific contexts. This appears to distinguish biology from physics, and suggests that explanation here is regulated by special principles, as it often is in the fragile sciences. Conversely, there are approaches within the social sciences in which individuals are treated as simple, unanalyzed elements in formal and statistical models used to predict large-scale behavior, much as atoms and molecules may be treated within the physical sciences. For example, game theoretical models of rationality and social change abstract away from individuals in this way. They posit a few simple psychological parameters, and use these to predict the outcomes of interactions between these scaled-down individuals.[17]

What of the term "fragile sciences" as the moniker for the cognitive, biological, and social sciences? Given that human agents are paradigmatic or representative individuals in these sciences, one might think that "human sciences" is descriptively more informative. But like "behavioral sciences" or "special sciences," this is a term whose connotations are more misleading than helpful, and whose extension differs from that of the range of sciences that I have in mind.

"The human sciences" denotes simply those sciences that attempt to understand human nature in one or more of its dimensions: biological, psychological, behavioral, social. The term thus suggests continuities between humanistic studies of human nature that precede the disciplinization of the psychological and social sciences in the nineteenth century and subsequent, disciplinary-based research. My overall concern is largely with conceptions of the individual in the contemporary biological, cognitive, and social sciences. Even if human agents are paradigmatic individuals in these sciences, their conceptualization in contemporary work seldom attempts to capture the essence of humanity, or to grapple with the loftier goals of earlier inquiries, such as "Man's place in nature." The conceptualization of the individual has become more partial and less encompassing, but also more closely tied to models, techniques, and research strategies in particular sciences.[18]

One reason to coin a term, rather than to make do with an existing one, such as "the human sciences" or "the behavioral sciences," that is particularly relevant for *Genes and the Agents of Life* is that the greater part of the biological sciences have more to say about nonhuman than about human agency, and about nonbehavioral than behavioral aspects of living agency. As I intimated earlier, human beings constitute just one special object of study within zoology. If "the human sciences" is taken, as it often is, to refer to psychology plus the social sciences, then we need a term that refers to these plus the biological sciences.

But why *fragile* sciences? Two brief reasons. The first is that it both parodies and transcends traditional divisions between the sciences: between the natural and social sciences, hard and soft science, and the physical and human sciences. These dichotomies too often drive views of the nature of science, and inevitably privilege the natural, the hard, and the physical over the social, the soft, and the human. "Fragile sciences" serves as a partial counter to these tendencies that pervade not only philosophy but also education, popular culture, and science itself. The second reason for "fragile sciences" is the cluster of ideas that fragility calls to mind. Fragile things can be easily broken, are often delicate and admirable in their own right, and their labeling as such carries with it its own warning, which we sometimes make explicit: handle with care! But they are also both strong and weak at the same time, and their fragility lies both in their underlying physical bases and in how it is that we treat them. No doubt, "fragile sciences" triggers other meanings, and should two reasons not be reason enough, let that be a third.

8. A PATH THROUGH *GENES AND THE AGENTS OF LIFE*

This book, like its companion *Boundaries of the Mind*, has four parts. Each part focuses on one key biological agent and a selection of the issues to which that agent is central.

In Part Two, I shall discuss organisms, paradigms of both living agents and of biological natural kinds. It is here that I shall return to explore some aspects of the notion of agency introduced in this chapter, and take up the challenges to the Aristotelian view of species posed by both the species-as-individuals thesis and pluralism about species concepts.

The central biological agent in Part Three is that of the gene, and there I will be concerned particularly with genetics and developmental biology and the relationships between them. The last decade or so has seen an explosion of work in both fields. In genetics, the Human Genome Project has provided biology with its most extensive (and expensive) instance of Big Science to date. In its name many striking discoveries about genes and how they operate have been made. Even more striking claims about what genes can do and their role in medicine, disease, and organismic development regularly pepper our newspapers. Genetics has, of course, contributed to developmental biology, but developmental biology itself has recently taken several directions that ascribe a more limited role to genes as the agents of organismic development than one might have expected from taking many of the claims made about genes at face value. Here I will be concerned with the conceptions of genes and development in play, including the metaphors used to describe both, and with the significance of so-called developmental systems theory for our conception of genetic and developmental agency.

In Part Four I focus on groups as putative evolutionary agents, returning to the issue of the levels of selection raised earlier in this chapter. As I implied in that discussion, "group selection" has been somewhat of a catch-all phrase for the operation of natural selection on units larger than the individual organism. One of our central tasks here will be to articulate group selection in a way that makes sense of this diversity. It is here that I will turn most explicitly, and most critically, to the pair of themes that I sketched in section 3 – that of the hierarchical structure to, and the pluralistic nature of, the biological world. I shall consider the most widely discussed example of group selection in the wild, that of the evolution of the myxoma virus in populations of rabbits in Australia since the 1950s, and argue that there are strong reasons to think that at least it, and by implication, other putative cases of group selection, can be adequately characterized without positing groups as agents of selection. While this suggests prima facie support for the recent pluralistic consensus about the levels of selection, the appearances here are deceiving. There is reason to reflect further on the ubiquitous view that the biological world is hierarchically structured.

Thinking about Biological Agents

1. TOOLS OF THE TRADE

I have suggested that we think of biology and biological agents in the broader context of the fragile sciences. One reason is that there is a useful conceptual toolkit – of concepts, distinctions, theses, and claims – that can be readily applied to the conception of individuals and agents across the cognitive, biological, and social sciences.

In this chapter, I introduce and illustrate some of the concepts and claims that will play central roles in the substantive discussions of biological agency in the remainder of the book. Most of these are likely to be novel for biologists, and they have not been extensively treated within the philosophy of biology. Their real import for thinking about biological agents lies in the substantive discussions in Parts Two through Four.

What are these tools of the trade? They are: the debate between individualists and externalists within the cognitive and biological sciences; the methodologies that accompany hierarchical thinking in the biological sciences; the notions of taxonomy, realization, and determination; the thesis of smallism as a general metaphysical view guiding research in biology; and, returning to the idea that life and mind are at least sometimes fruitfully considered together, the claim that a cognitive metaphor often operates in the biological and social sciences, particularly in its characterization of biological agents.

These concepts and claims constitute a motley crew. While there is no simple way in which they are related, I hope to trace several threads connecting them throughout the remainder of *Genes and the Agents of Life*. Here I aim simply to introduce them and to provide some idea of their potential relevance to our thinking about organisms, about genes, and about groups in the biological world.

2. THREE FORMS OF INDIVIDUALISM IN BIOLOGY

One way in which the role of the individual has been made prominent in the fragile sciences is via the defense of one or another form of individualism. In psychology and the cognitive sciences more generally, individualism is the thesis that psychological states should be construed without reference to anything beyond the boundary of the individual who has those states. Put loosely, for the purposes of scientifically understanding the mind, the individual is the boundary for cognition. Minds are located inside individuals, and we need not venture beyond the boundary of the skin in characterizing any individual's mind. Thus, if one is an individualist about the mind, then one should abstract away from an individual's environment in characterizing her psychological states.

A more precise expression of individualism says that psychological states should be taxonomized to supervene on the intrinsic, physical states of the individuals who instantiate those states. (A property, A, supervenes on another, B, just if no two entities can differ with respect to A without also differing with respect to B.) This is usually taken to mean that if two individuals are physically identical, then they must also be *psychologically* identical. It is for this reason that individualism about psychology has often been presented as a view that follows from the acceptance of materialism or physicalism about the mind and viewed as an instance of a quite general constraint on scientific taxonomies.

In *Boundaries of the Mind* I discussed individualism about the mind in some detail. There I argued that there were three essential characteristics of this view of cognition that were also shared by individualistic theses in the biological sciences. I shall briefly summarize generalized versions of these, and show how we can use this characterization of individualism to analyze a debate within biology that we discussed in Chapter 1, that concerning the agents of selection.

First, individualism is a normative thesis about how one ought to conduct our science. In the case of cognition, it proscribes certain views of our psychological nature – those that are not individualistic – and so imposes a putative constraint on the sciences of cognition. Second, this constraint itself is claimed to derive either from general canons governing science and explanation or from entrenched assumptions about the nature of the subject matter of some particular science. It meshes with existing explanatory practices that have met with considerable success in the past, and is an empirical rather than an *a priori* constraint on the sciences of the mind. Third, approaches to science that are not individualistic are both methodologically and metaphysically misguided. In cognitive science, they go methodologically awry in that the most perspicuous examples of explanatorily insightful research paradigms for cognition – computational approaches – have been individualistic. Rejecting individualism, and thus such approaches, leaves one in methodological limbo. They go metaphysically awry in a corresponding way, relinquishing the insights into mental causation afforded by computational views of cognition. Along the same lines, rejecting individualism about the mind appears to treat the cognitive sciences in a special way, suggesting that its taxonomic categories are exempt from constraints that apply to the rest of nature.

In short, individualism about a particular science is a normative, empirically grounded constraint that guides how that science is practiced. In the biological sciences, we can readily see individualism as an issue that arises in the debate over the agents of natural selection. Here individualism is the view that the organism is the largest unit on which natural selection operates. Thus, proponents of genic selection, who claim that natural selection can always be adequately represented as operating on genes or small genetic fragments, are individualists about the units of selection, as are those who adopt the traditional Darwinian view that allows for only organism-level selection. To embrace higher levels of selection, such as group selection, is to reject individualism. Like individualism in psychology, individualism about the agents of selection is a putative, normative constraint that derives from existing explanatory practice and whose violation, according to its proponents, involves both methodological and metaphysical mistakes.

First, individualism about the agents of selection implies that individual organisms act as a boundary beyond which evolutionary biologists need not venture when considering the nature of what it is that competes and is subject to natural selection, and thus evolutionary change. Second, by focusing on the individual and what lies within it, one can best understand the dynamics of adaptive change within populations of organisms, whether it be via population genetic models, through the deployment of evolutionary game theory, or by means of the discovery of the forms that individualistic selection takes. This constraint on how to think about the operation of natural selection builds on the specific explanatory successes of models of kin selection, reciprocal altruism, and other processes that articulate strategies that individuals might adopt in order to maximize their reproductive success. Third, flouting individualism creates both methodological and metaphysical problems avoided by individualistic approaches. Methodologically: since individualistic models of selection, such as kin selection or reciprocal altruism, putatively explain the full range of observed behaviors in evolutionary terms, to reject individualism is to abandon real explanatory achievement. Metaphysically: just how does natural selection transcend the level of the individual and go to work directly on groups? A common response to claims of group selection is either that the appropriate model of group selection really boils down to a variant on an existing individualistic model of selection, or that it requires assumptions that rarely hold in the actual world. In the former case, we simply have individualism by another name; in the latter, our models are mere models and fail to correspond to how the world actually is.

My argument in *Boundaries of the Mind* contained a negative and a positive strand concerning individualism about the mind. Although there are strong prima facie reasons to think that individualism is a constraint on the sciences of cognition, and much cognitive science has been individualistic, in fact individualism should be rejected. More positively, there are contrasting *externalist* views within the cognitive sciences worth developing further. These require rethinking many concepts central to the philosophy of mind and cognitive science, such as physicalism, computation, and representation.

Whether individualism imposes a constraint in other areas of the fragile sciences remains a substantive issue. While my claims about cognition are not directly relevant to individualism about biology, they at least caution against an unreflective acceptance of individualism. The plausibility of individualism more generally depends in part on how the debate over individualism is generalized or transformed in moving from the cognitive to the remainder of the fragile sciences. There are at least three ways of doing so.

The first, exemplified by the previous example of the debate over the agents of selection, is simply to transpose the issue of whether the individual organism serves as a boundary of some kind that constrains the form of the corresponding science (see Table 2.1). This form of generalization takes a bird's-eye view of the individualism debate over the mind, abstracting away from the metaphysical complexities – expressed in terms of realization, determination, and intrinsic properties, for example – that talk of supervenience brings in its wake.

TABLE 2.1. *Individualism: cognition and biology*

	Cognitive	Biological
Individualistic constraint	Individual as a boundary for cognition	Individuals as largest unit acted on by selection
Empirically grounded explanatory practice	Appeal to scientific taxonomy in general or to specific features of psychology (e.g., computational theory of mind)	Principle of parsimony; success of individualistic theories, such as kin selection, reciprocal altruism
Denials:		
Methodological:	Abandoning best research programs	Abandoning best research programs
Metaphysical:	Giving up on mental causation and mechanisms	Group selection mysterious

It is prima facie extremely plausible to think that at least some areas within the biological sciences, such as physiology, genetics, and developmental biology, are individualistic in this sense. Such subdisciplines are concerned with units that are parts of, sometimes very small parts of, individual organisms. Since the systems that such units constitute can be understood in abstraction from much of the rest of the body of the organism, it is difficult to see why the same should not be true of the world beyond the individual. So at least these parts of biology would seem to be individualistic.

The second way to extend the debate over individualism from the cognitive to the biological domain is to adapt the supervenience formulation of individualism and explore whether *biological* properties supervene on the intrinsic, physical properties of individuals. Although I implied previously that one motivation for individualism in psychology was the idea that individualism held more generally in the sciences, there are many examples of biological properties that prima facie flout the constraint of individualism when expressed in terms of this supervenience claim. (In fact, such examples have provided one basis for challenging the presumption of individualism about the mind.) These examples include evolutionary fitness, being highly specialized, and being a predator, all properties of individual organisms or species, as well as properties of phenotypic traits or behaviors of individuals, such as being an adaptation, a homology, or a spandrel. What all of these biological properties share is that they are *relational* or *contextual*, such that something can gain or lose the property simply through a change in that thing's relations, or in the context in which it exists.

To take one of these examples, whether a given trait, such as a wing of a particular shape and structure, is an adaptation (say, for flight) depends in part on the history of that structure. When evolutionary biologists argue about whether a wing-shaped structure is an adaptation for flight, this sort of historical consideration is paramount, and the lineage history is itself individuative. History is not simply an epistemic clue to discovering something intrinsic to the organism or trait itself, but part of what makes a given structure an adaptation for flight. Without the right kind of history, wing-shaped structures are not, cannot be, adaptations for flight, no matter what else is true of them, any more than pieces of paper lacking the right kind of history can be dollar bills. Two pieces of paper that were not just indistinguishable but identical in their intrinsic, physical properties could differ with respect to the property of being a dollar bill simply because only one of them was produced by the government treasury, the other by your neighbor. Likewise, of two wing-shaped structures identical in their here-and-now physical features, only the one that resulted from a lineage in which flying had been naturally selected would be counted as an adaptation for flight.

It is common for the biological sciences to explore properties that metaphysically depend on more than the intrinsic, physical features of the agents that have them. This is just to say that many biological properties do not supervene on the intrinsic, physical properties of the agent that has the property. This creates a hiatus between biology and the here-and-now because historical and other relational differences between organisms and their structures need not be mirrored in the intrinsic, physical properties that each has. Thus, relational properties can metaphysically determine what kinds and generalizations organisms and their structures are subsumed under, and in principle we could have physically identical structures that, as in the case of our wing-shaped structures, are subject to distinct categories and generalizations.

If such relational properties are widespread in the biological sciences, particularly in the evolutionary and ecological aspects of biology, then prima facie biology is not individualistic in this second sense. But to move beyond the appearances we have to return to some of the metaphysical complexities bypassed by the first construal of the individualism-externalism debate. In particular, we need to attend to the nature of relational properties (and the contrast between them and intrinsic properties), and philosophical notions foreign to the ear of many a biologist, such as determination and realization. In doing so, we may have reason to reconsider the apparent individualism of physiology, genetics, and development, and to explore the relationship between our two senses of individualism.

In considering whether biological properties supervene on an individual's intrinsic, physical properties we have maintained a focus on our paradigm individuals, human agents and organisms. The third way to transpose the individualism issue from the domain of cognition to that of biology drops this focus, and generalizes on the notion of an agent or individual. Thus, we move from our paradigm individuals, organisms, to other kinds of things that are in some sense treated as individuals. As we have already seen, in the biological sciences these include groups and species, but also living things that are contained within paradigmatic individuals, such as bodily organs and obligate parasites, and other biological entities, such as pathways and systems. Such biological agents are often construed as individuals in their own right.

Individualism in the biological sciences on this third construal is the view that biological taxonomy at *any* level is by the intrinsic, physical properties of the entities taxonomized. We might express this view as holding that what makes any biological entity the kind of thing it is are facts about what is inside its boundary, what it is physically constituted by, or what causal powers it possesses. For example, what makes something a gene (or a specific type of gene), or a protein, or a heart, are facts about that thing's constitution or the causal powers that it has. This version of individualism will be particularly relevant as we move "up" and "down" from the organism to examine other kinds of biological agents. Table 2.2 provides a summary of these three ways to construe individualism within the biological sciences.

Keeping the parallel between the cognitive and biological sciences, I shall call those who deny one of these individualistic positions within biology *externalists*. Thus, proponents of group selection are externalists, and the name is descriptively apt in that they appeal to factors external to organisms – population density and structure, dispersal mechanisms, resource limitations – in identifying the agency of natural selection.

In denying one or another form of individualism, externalists are not, however, proposing their own version of a normative constraint on the corresponding science. This is in part because they adopt a more pluralistic view of scientific taxonomy, one that allows for a place for causal

TABLE 2.2. *Three construals of individualism in biology*

Construal	Characterization	Sample Issue
Individual as a boundary	Can bracket off or ignore world beyond the individual in doing biological science	Are physiology, genetics, and developmental biology individualistic?
Biological properties as supervenient on the organism	Intrinsic physical properties of organisms constitute a supervenience or realization base for biology	What is the significance of the prevalence of relational and historical properties in ecology and evolutionary biology?
Biological properties of X as supervenient on what's inside X	A generalization of the second construal, from organisms to individuals more generally	Do intrinsic, causal powers play a special role in characterizing biological kinds and taxonomies?

powers and intrinsic properties, but that also recognizes a taxonomic or individuative role for the relational and historical properties that individuals possess. Externalists are likely to view scientific taxonomies and scientific explanation as sensitive to a range of factors, and to be skeptical about the prospects for recipe-like prescriptions or generalizations regarding proper scientific taxonomy of the sort that individualists propose. Thus, a consideration of the positive forms that externalism takes will return us to the issue of pluralism in the biological sciences.[1]

3. TAXONOMY AND TAXONOMIC PRACTICES

In characterizing individualism, I have mentioned taxonomy in the biological sciences, and I want now to head off one misunderstanding of what I mean by this. In biology, "taxonomy" (or "systematics") is the name given to a particular area of evolutionary biology that focuses on the correct way of classifying organisms into natural kinds, and the relationships between those kinds. So construed, taxonomy is sometimes caricatured as a somewhat dusty and dated area of the biological sciences, one with more affinity to nature lovers and amateur biologists than to the technology and experimental savvy of professional biology. It concerns nomenclature more than methodology, how we conceptualize relatively easy-to-observe biological units rather than the deep nature of the biological world per se, and what is done after the scientific work is in, rather than a part of that work itself. It takes place in museums, not in laboratories.

This caricature of taxonomy is mistaken on nearly every single point. Taxonomic practices in biology range from carbon dating and anatomical reconstruction of fossils in paleozoology to molecular phylogenetic analysis of both microorganismic and plant lineages, and range up and down various biological hierarchies. But my main point here is that when I talk of individualistic or externalist taxonomic practices I mean something much more general than what happens within the subarea of systematics. Taxonomies are simply categories and kinds, and they exist throughout the biological sciences. Taxonomic categories in genetics and developmental biology include cells, genes, tissues, organs, bones, cytoskeletal structures, membranes, pathways, modules, molecules, and various biological systems (nervous, digestive, immune, respiratory), as well as their parts and the determinate forms they take. In ecology they include organisms, demes, adaptive niches, ecosystems, communities, symbiotic pairs, and predatory-prey systems.

Each of these kinds of thing is a putative, biological agent. One way to think of the debate between individualists and externalists in biology is in terms of whether there are normative, empirically based constraints on the nature of biological agents. In particular, do our paradigm biological agents – organisms – serve as some sort of boundary for characterizing biological agency more generally? Putting this more precisely (and taking up the second sense of individualism): do these kinds of biological agents and the properties they have supervene on what lies inside the boundary of organisms? Or, to move to our third form of individualism: do the properties of all biological agents whatsoever supervene on the intrinsic, physical properties of those agents? To answer "yes" to one of these questions is to accept a form of individualism about biology. To answer "no" is to embrace the corresponding form of externalism.

4. THE ART OF HIERARCHICAL THINKING IN BIOLOGY

In Chapter 1, I noted that both organisms and hierarchies feature prominently in how we conceptualize the natural world, particularly in thinking about the living world. Formal training in biology brings with it two techniques for moving beyond the organism as the focus of biological inquiry, either decomposing the organism into its constituents – genes, cells, tissues, organs, bodily systems – or integrating it into larger biological agents through the relationships it features in – with other conspecific individuals, their ancestors, local environments, whole ecosystems. In this section, I shall make some brief comments about these two methodologies, what I call *constitutive decomposition* and *integrative synthesis*, and their significance for thinking about agency.

In constitutive decomposition, one begins with an entity and what it does and then decomposes it and its functions into smaller entities with more specific functions. This explanatory strategy has been used particularly effectively in physiology, where one begins with a function (for example, blood circulation), identifies the system that performs it (circulatory system), and details the parts of the system (blood, venules, arteries, heart) that perform the subsidiary functions. Importantly, constitutive decomposition can be applied recursively at each new level revealed by its preceding application, uncovering further details of the mechanisms through which bodily functions are realized or instantiated.

This research strategy has been pervasive within the cognitive sciences, where it is often called *homuncular functionalism.* "Functionalism," because of its focus on mental functions, whether they be the classic, personal-level categories of perception, memory, volition, and action, or more recently posited and subsidiary functions, such as 3-D image construction from 2-D images in perception, or storage techniques for short-term memorization. And "homuncular," because the strategy can be thought of as beginning with an agent, a "little person," that performs the function, which is then replaced through application of the strategy by two or more further agents that perform the more specific functions posited. Daniel Dennett has eloquently shown the importance of this methodology within artificial intelligence, and William Lycan has suggested the fruitfulness of conceiving of psychological individuals as *corporate entities* constituted by subpersonal agencies of the type identified in constitutive decomposition.[2]

Within the biological sciences, constitutive decomposition has been championed both as a methodology for integration with the physical sciences and as a reductionist research strategy aimed at uncovering the mechanisms underlying biological processes. Over the past fifty years, the tendency to constitutively decompose organisms and the processes in which they function and concentrate on their smaller parts – on amino acid sequences in protein formation in biochemistry, on microtubule growth patterns in cell biology, or on genome sequencing in genetics – has grown markedly within biology. Sometimes constitutive decomposition results in the organism disappearing altogether from view. What begins, for example, as a disease of the person becomes a malfunctioning of some bodily system, which is then understood as a breakdown of cellular metabolism: from cancer to lymph node blockage to runaway cell division.

The alternative strategy, integrative synthesis, involves identifying the larger system of which an organism with a given property is a part. Integrative synthesis is more prevalent in ecology and evolutionary biology, and it provides one way to understand and study the sort of relational properties prevalent in those areas of biology that I mentioned briefly in the previous section. Like constitutive decomposition, integrative synthesis can be applied at any level in a biological hierarchy of agents. Often enough, its application will move one from a given agent to another.

It may be tempting to think of these methodological strategies simply as inverses of one another. Yet this would be to overlook a significant difference between them. For like individualism and externalism, the methodologies of constitutive decomposition and integrative synthesis carry with them quite distinct metaphysical views of agents and their properties. In the next few sections, I shall try to spell out these views more precisely by introducing the technical philosophical notions of realization and determination, beginning with some links between methodology and metaphysics.

5. METHODOLOGY AND METAPHYSICS

While talk of "realization" and "determination" is alien to biologists and scientists more generally, these notions are crucial for understanding the relationship between putatively distinct hierarchical levels in science. Determination is perhaps most easily conceptualized diachronically, and I want to introduce this notion of determination before moving to the synchronic notion of determination relevant to characterizing realization.

Causation is often thought of as a diachronic relation. Putting aside probabilistic causation, it is a relationship of determination between (say) processes or events related in time, such that an earlier process or event – a bacterial infection – causes a later process or event – a festering of the site of infection. Causation is a relation of determination in that the occurrence of the cause fixes or determines the occurrence of the effect, given that certain conditions remain in place over time.

Realization, by contrast, is a relation that holds at a time, and it is typically thought of as holding between entities "at different levels." Talk of realization has its home in thinking about the relationship between the mental and the physical. A common expression of physicalism about the mind says that the mind is realized in the physical constituents of the body. Likewise, with particular mental states, such as Ted's having a particular pain at a given time, which might be realized in Ted's nociceptive system being in a particular physical condition at that time. Realization is a relation of determination at a time in much the sense that causation is a relation of determination across times: the presence of the realizing state, property, or process fixes or determines the presence of what it realizes.

For this reason, we might naturally talk of a range of biological agents and the properties they have in terms of the notion of realization. For example, agents such as genes are realized by segments of DNA, and metabolic functions are realized by the activities of agents such as distinct biochemical pathways. Organisms have a physical realization, as do their biological properties, although if many of these properties are not individualistic, then they will be realized by more than simply the intrinsic, physical properties of the agents that have them.

In general, properties (features, characteristics, traits) are abstract features of entities that are realized or instantiated in particular physical arrangements of matter. Often enough, these realizations are fully contained within the boundary of the entity that has the property. In such cases, we have what I call an *entity-bounded* realization. A knife, for example, may have the property of being sharp, and that property of the knife is realized in the particular character and arrangement of the physical constituents of the knife. The sharpness of the knife has an entity-bounded realization. A chemical substance may have the property of being in a liquid state, and again this property is one that has an entity-bounded realization.

The connection between the research strategy of constitutive decomposition and a metaphysics that posits entity-bounded realizations should be clear. Suppose that many of the properties possessed by entities have entity-bounded realizations. Since these realizations involve the smaller constituents of those entities, we should explain those properties in terms of the properties of these constituents. Hence, the need for constitutive decomposition.

But not all properties have entity-bounded realizations. Some have what I call a *wide* realization. Wide realizations also involve the arrangement of physical entities, but they are not contained within the physical boundary of the individual who has the corresponding property. They extend beyond the individual. Just as entity-bounded realizations are

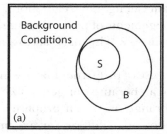 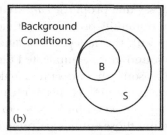

FIGURE 2.1. (a) Constitutive decomposition, involving entity-bounded realization, and (b) integrative synthesis, involving wide realization

uncovered through constitutive decomposition, wide realizations are revealed through integrative synthesis. And just as an entity-bounded realization is contained within some system (for example, respiratory, cardiovascular) that is a constitutive part of the individual, a wide realization is contained, in turn, within some system (for example, predator-prey, ecological) that the individual constitutes.

We can make this parallel between these two forms of realization and our explanatory strategies more vivid in a simple figure. Let B be the subject or individual *b*earer of some higher-level, biological property, P, and S be the *s*ystem in which that property is realized. In constitutive decomposition, S is a part of B; in integrative synthesis, B is a part of S (see Figure 2.1). As I have said, the metaphysical and explanatory framework depicted in Figure 2.1a is exemplified within physiology, where P might be the property of having stomach pain, B the individual with the pain, and S the digestive system in a particular physical state (for example, being extremely full of food). Since S is contained in B, P here has an entity-bounded realization. By contrast, the framework encapsulated in Figure 2.1b is exemplified within ecology, where P might be the property of being a predator, B is the predatory organism itself, and S is the particular predator-prey system of which that individual is a part. Since B is contained in S, P here has a wide realization.

6. PROPERTIES, REALIZATION, AND DETERMINATION

Given that biology is often nonindividualistic in one or more of the three senses that I have introduced, the biological sciences are rife with properties that have each of these kinds of realization. There are, however, a few further questions one might have about this talk of realization, especially if one is not a philosopher. I shall focus on just one cluster of questions.

Why not say that properties *are* what I am calling their realizations? Why not simply identify properties with arrangements of physical matter, and be done with this fancy talk of "realizations"? Why make all this metaphysics sound more complicated than it is?

One reason is that we want to make sense of properties that are shared by different entities. This is not least of all because our generalizations, including our scientific laws, seem to range over such properties and entities. But those entities are often composed of different arrangements of physical entities, or even arrangements of different physical entities. Since identity is transitive – if A is identical to B, and B identical to C, then A is identical to C – identifying properties with realizations would entail either identifying the realizations or distinguishing the properties that, at least at an intuitive level, seem identical to one another.

For example, we sometimes want to say that two bottles filled with gases have the same pressure, that is, the very same property, and that their sharing this property is causally responsible for various other facts about them (for example, their exploding or not exploding). Yet those bottles and those gases could have very different constituents, one made of aluminum and containing oxygen, the other made of steel and containing carbon dioxide. If the determinate pressure that the first had is strictly identical to what I have been calling its entity-bounded realization, and that of the second is strictly identical to *its* entity-bounded realization, then these entity-bounded realizations are, in turn, identical. Or, alternatively, there are really two distinct properties that the bottles filled with gas have, lumped together under the predicate of (say) contains a gas that exerts 10 kilograms of pressure per cubic meter. Given the physical distinctness of each bottle with respect to the other, the first of these options is implausible. And given that the generalizations we make about how the bottles behave presuppose their having the same property, the second of these options is likewise implausible.

This line of thought has been particularly poignant when the entity-bounded realizations have been very different from one another. Those working in the philosophy of mind and the cognitive sciences have often thought of this as the standard situation with mental properties and have described those properties as having *multiple realizations*. The realization of pain in humans and other actual creatures, such as octupuses, or merely possible creatures, such as silicon-based Martians, have been standard examples here. I do not want to enter the debate over whether such a view of mental properties and functions is justified, except to say that the differential realization that I have appealed to above need not carry all of the metaphysical baggage that the notion of multiple realization does.[3]

One piece of baggage that it does carry, however, one that the relation of realization shares with identity, is that realizations are *metaphysically determinative* of the properties they realize. I have already said that metaphysical determination is synchronic – it holds between two entities or sets of properties at a time – and that it contrasts in this respect with diachronic relations of determination, such as causation, that hold across temporal intervals. In *Boundaries of the Mind* I characterized the determinative nature of realization as a thesis, the *metaphysical sufficiency thesis*, that was widely endorsed in the philosophy of mind. It amounts to the claim that once the physical realization of a property is specified or (as philosophers say) "fixed," so too is the property itself: it is not possible to have the realization present without the property it realizes also being present, in a given case. If being a liquid is realized in a given case by a certain arrangement of physical molecules, then if just that arrangement of just those molecules is present in something, then that thing must have the property of being a liquid. Note, in light of the above point about multiple realization, that the converse is not true: it is possible to have something that is a liquid that has a different arrangement of physical molecules. Thus, the relationship of metaphysical determination involved in realization, unlike that in identity, is asymmetrical.

I have said that biologists themselves seldom talk of realization, yet such talk provides a natural way to express the relationships between agents in biological hierarchies. Although it is common to use the simple "is" and "are" of identity in order to express the relation between such agents, these expressions are most plausibly (and charitably, if we aim at truth in interpretation) viewed as short-hand for the more unwieldy "are realized by" or "is a realization of." Genes are claimed to *be* sequences of DNA, and a population of viruses on a rabbit's head to be nothing more than individual viruses located on a common host. But since the very same kind of gene or population can exist despite there being changes in the precise sequence of DNA or the specific individuals in the population, these are at most relations of realization, not of strict identity.

When individualism is expressed in terms of the notion of supervenience, it specifies a relation of metaphysical determination, for supervenience is determinative. Applied to organisms, individualism says that the intrinsic, physical properties of organisms determine its biological properties. Generalized to all kinds of biological agents, individualism entails a claim about what metaphysically determines all biological properties: the intrinsic, physical properties of agents. Were either form of individualism true, then what is inside the physical boundary of biological agents would deserve special attention, for it is here that we will find the properties that determine all other features of those agents. Despite the prima facie externalist character of much biological science, individualism remains a tempting view of large tracts of biology. One reason for this is the relationship of individualism to broader, influential metaphysical views.

7. SMALLISM AND BIOLOGY

It has seemed to me for some time that when it comes to metaphysics, there is a deep philosophical and scientific bias in favor of the small and so against the not so small. In keeping with times of political correctness, I have referred to this form of discrimination as *smallism*. Small things and their properties are seen as ontologically prior to the not-so-small things that they constitute. Smallism in the philosophy of mind is manifest in a near-exclusive focus on entity-bounded realizations for mental states. In the opening chapter of *Boundaries of the Mind*, I suggested that there is an affinity between individualistic views of the mind and smallism more generally. In this section, I shall provide some examples of smallist views from the history of biology, using them to illustrate the relations between smallism, constitutive decomposition, and individualism in the biological sciences.

By the end of the nineteenth century, constitutive decomposition had become the predominant methodology for thinking about organisms and other biological agents. The marine zoologist William Emerson Ritter considered this predominance a bias, and his two-volume *The Unity of the Organism*, published in 1919, was written largely in order to redress an imbalance that he saw as having arisen between this type of view, what he called *elementalism*, and *organismalism* in the biological sciences. Ritter's basic goal was to show that it was only by considering an organism's parts in relation to the whole organism that one could truly understand biological functions and processes. Ritter's efforts to one side, it would not be inaccurate to characterize the twentieth century as the triumph of smallism in the biological sciences.

Consider the Nobel laureate François Jacob's summary of contemporary biology in his elegant history of heredity, *The Logic of Life*: "The aim of modern biology is to interpret the properties of the organism by the structure of its constituent molecules. In this sense, modern biology belongs to the new age of mechanism." In fact, Jacob's overall view of the history of heredity depicts that history as the steady replacement of the not-so-small with the small. Jacob is, again, his own best spokesperson:

There is not one single organization of the living, but a series of organizations fitted into one another like nests of boxes or Russian dolls. Within each, another is hidden. Beyond each structure accessible to investigation, another structure of a higher order is revealed, integrating the first and giving it its properties. The second can only be reached by upsetting the first, by decomposing the organism and recomposing it according to other laws.[4]

Jacob's general views frame his equally memorable characterization of the place of the metaphor of a program in describing the nature of heredity. He says,

What are transmitted from generation to generation are the 'instructions' specifying the molecular structures: the architectural plans of the future organism. They are also the means of executing these plans and of coordinating the activities of the system. In the chromosomes received from its parents, each egg therefore contains its entire future: the stages of its development, the shape and the properties of the living being that will emerge. The organism thus becomes the realization of a programme prescribed by its heredity.[5]

Organisms thus become genetic programs, programs inscribed in the sequence of base pairs that constitute the genome. In Jacob's view, over the past few hundred years a focus on "arrangements of visible surfaces" was replaced successively by cells, genes, and nucleic acids in scientific attempts to understand the familiar and age-old phenomenon of the generation of like from like in the biological world. The operation of the smallest of these entities, nucleic acids, is to be understood in terms of a computational paradigm, one that makes it seem that all hereditary agency is located in the "program" itself – smallism, and individualism (in all three senses), with a vengeance.

The rise of the idea that cells are the fundamental units of life, those out of which all of life is constructed, was important not only for theories of hereditary transmission but for the biological sciences more generally. As the cytologist E. B. Wilson put it at the outset of his influential text *The Cell in Development and Inheritance*:

During the half-century that has elapsed since the enunciation of the cell-theory by Schleiden and Schwann, in 1838–39, it has become ever more clearly apparent that the key to all ultimate biological problems must, in the last analysis, be sought in the cell.[6]

The Schwann-Schleiden theory played a critical role in, amongst other things, Claude Bernard's development of physiology as a distinct field from anatomy; in articulating a view of life that not only unified the plant and animal kingdoms but which also revealed their relationship to the protista (characterized as "unicellular life"); and in providing the basis for explaining organismic development. Every living thing was composed of cells, early life was unicellular, spawning later multicellular life, and organismal development proceeded through the division of these basic biological units, cells.

The placement of the cell at centerstage in a range of biological debates thus gave some naturalistic bite to the traditional idea that the living world was hierarchically organized, from the small (cells) to the not-so-small things (organs, bodily systems, organisms) that they physically compose. Cells were a fundamental (even if not ultimate) unit in this hierarchy, and their study was basic to understanding the workings of the living world. In principle, researchers can work down such a hierarchy, seeking to explain the operation of higher-level units in terms of the properties of and relations between lower-level units, or upward, beginning with lower-level units and trying to show just how they causally explain a higher-level property or process.

Given this characterization, constituent decomposition and integrative synthesis seem like complementary strategies of explanation. I have already cautioned, however, against thinking of these two strategies, the decompositional and integrative, simply as two sides of the same coin. This symmetry between "up" and "down" strategies of explanation is compatible with – indeed, facilitates – smallism. For in both explanatory directions the metaphysical and explanatory burden lies very much with the small, rather than with the not-so-small. It is properties of entities specified at lower levels that are invoked, whether we look up from the small to the not-so-small, or back the other way.

Such a view is common in contemporary discussions, even those that present a relatively ecumenical view of decomposition and integration. For example, consider William Bechtel and Robert Richardson's fascinating study of strategies of localization in nineteenth- and twentieth-century neurophysiology and biochemistry, *Discovering Complexity*. Bechtel and Richardson analyze a range of examples from the history of the psychological and biological sciences with an eye to highlighting the role and nature of strategies of decomposition and localization. In their view, when strategies of "upward" integration attempt to challenge the smallist orientations – by appealing, for example, to emergent, holistic, or whole-greater-than-the-sum properties – they must content themselves with offering a more accurate *description* of the phenomena to be explained, rather than with providing an alternative explanation of the phenomena of interest. This limits the appeal of such views within the relevant scientific community. Labeling integrative synthesis as antimechanistic, Bechtel and Richardson put it thus: "Lacking a systematic program of research, the antimechanistic program is virtually guaranteed to fail in gaining adherents."[7]

This sort of asymmetry between the small and the not-so-small is also manifest in George C. Williams's defense of smallism – what he calls "reductionism" – in evolutionary biology. As one of the architects of the selfish gene view of natural selection, Williams had earlier argued for a constrained view of how natural selection operated, one that eschewed group selection as anything more than a conceptual possibility. For Williams, genes are likely to be the only widely applicable biological device for conducting the sort of "bookkeeping" that he sees as central to detecting, measuring, and predicting evolutionary change. (Williams also allows that *gene pools*, found in large-scale populations, can be such devices, but views their use as more limited.) In Williams's view, explaining the not-so-small in terms of their component, small parts and their properties is the only general research strategy that has led to scientific progress, even if one needs to combine it with purely historical explanations to arrive at a full understanding of any particular biological phenomenon. Like Jacob, Williams recognizes the existence of alternative "holistic" or "emergentist" approaches to understanding the living world; like Bechtel and Richardson, he finds little empirical solace in this. Of those who would resist his smallist view of evolutionary biology, he says:

The history of science shows the inevitability of widespread emotional dissatisfaction with reductionist interpretations of the world. I suppose there are people who think that the theory of sexual selection robs the grossbeak's song of its music or that kin selection makes a mockery of brotherly love. There used to be those who thought that Newtonian optics destroyed the beauty of the rainbow ... [8]

One question that the remainder of *Genes and the Agents of Life* can be seen as addressing is whether the rejection of this sort of reductionism is merely emotional dissatisfaction.

8. THE COGNITIVE METAPHOR

The question of the place of metaphor in scientific explanation has received only passing attention from philosophers and scientists. The geneticist and evolutionary biologist Richard Lewontin has ascribed metaphor a central role within scientific explanation and theorizing. At the outset of his *The Triple Helix*, he says,

It is not possible to do the work of science without using a language that is filled with metaphors. Virtually the entire body of modern science is an attempt to explain phenomena that cannot be experienced directly by human beings, by reference to forces and processes that we cannot perceive directly because they are too small, like molecules, or too vast, like the entire known universe, or the result of forces that our senses cannot detect, like electromagnetism, or the outcome of extremely complex interactions, like the coming into being of an individual organism from its conception as a fertilized egg. Such explanations, if they are to be not merely formal propositions, framed in an invented technical language, but are to appeal to the understanding of the world that we have gained through ordinary experience, must necessarily involve the use of metaphorical language. Physicists speak of 'waves' and 'particles' even though there is no medium in which those 'waves' move and no solidity to those 'particles.' Biologists speak of genes as 'blueprints' and DNA as 'information.'[9]

I shall have more to say about specific metaphors used in genetics and development in Part Three. Here I want to draw attention to one distinctive aspect of the use of metaphor in biology and the social sciences.

In Chapter 1, I briefly raised the issue of how life and mind were related. Some connections between the two are made through metaphor. In particular, in the biological and social sciences we often metaphorically extend our conception of mind from our paradigmatic individuals, human agents, to things that do not have minds. This use of the cognitive metaphor gains purchase across various biological hierarchies. In characterizing biological agents that are smaller than organisms, and typically a part of them, the cognitive metaphor is manifest in Richard Dawkins's metaphor of the selfish gene; in cell biology, where cells are described as recognizing, remembering, preferring, and seeking certain other cells or molecules; and in immunology, where the immune system is conceptualized as distinguishing self from nonself.[10]

The cognitive metaphor can also be found when we turn to biological agents larger than the organism, such as superorganisms, and groups of organisms more generally. It is pervasive in characterizations of nonbiological agents, such as social institutions and practices, or even whole societies or cultures. Each of these may be said to decide, choose, prefer, or remember, and, in at least many cases, this talk is metaphorical.

The evolutionary biologist David Sloan Wilson has revived the idea that groups have minds in the context of his extended defense of group selection. Noting that "[g]roup-level adaptations are usually studied in the context of physical activities such as resource utilization, predator defense, and so on," he says

groups can also evolve into adaptive units with respect to cognitive activities such as decision making, memory, and learning. As one example, decision making is a process that involves identifying a problem, imagining a number of alternative solutions, evaluating the alternatives, and making the final decision on how to behave. Each of these activities can be performed by an individual as a self-contained cognitive unit but might be performed even better by groups of individuals interacting in a coordinated fashion. At the extreme, groups might become so integrated and the contribution of any single member might become so partial that the group could literally be said to have a mind in a way that individuals do not, just as brains have a mind in a way that neurons do not.[11]

The anthropologist Mary Douglas has also explored the idea of group minds in the social sciences in her *How Institutions Think*, noting that the "very idea of a suprapersonal cognitive system stirs a deep sense of outrage." As I said in Chapter 1, individual human beings (and some animals) are privileged as cognitive agents, and that is surely one of the grounds for resistance to the idea that groups, biological or social, have minds.[12]

Given the prevalence of the cognitive metaphor across the biological and social sciences, the question is what we should make of these sorts of metaphorical attributions of psychological properties, states, and capacities. What is the status of the explanations that such attributions feature in? Does their ubiquity suggest some sort of lacuna or even defect in the fragile sciences?

Since I have discussed the application of the cognitive metaphor to agents larger than the individual organism in some detail in *Boundaries of the Mind*, here I will be particularly interested in its uses in characterizing agents smaller than organismic agents, such as genes and bodily systems. But, as we will see in the next few chapters, I also think that the cognitive metaphor operates behind the scenes in our conception of the organism as well. Life and mind are not as readily dissociated as one might expect.

PART TWO

SPECIES, ORGANISMS, AND BIOLOGICAL NATURAL KINDS

3

What Is an Organism?

1. THE FAMILIAR AND PUZZLING WORLD OF ORGANISMS

In the introductory chapter, I said that organisms were paradigms both of living agents and of biological natural kinds. They appear ubiquitously across hierarchically ordered lists of biological entities and are familiar both to common sense and within the biological sciences. When we think about the biological world, organisms leap out at us immediately as the agents of life to such an extent that it is sometimes difficult to envisage life without organisms.

So everybody knows what organisms are. Given all of this, it may seem unwise for me to waste all of our time with a chapter on the question "What is an organism?" But even if organisms are obvious, almost inevitable denizens of our thinking about life, the concept of an organism stands in need of some elucidation. Consider the following three examples and questions that naturally arise about them.[1]

In the early 1990s, a team of biologists led by Myron Smith reported in the journal *Nature* that they had found that fungus samples of the species *Armillaria bulbosa* taken over a region of fifteen hectares in Michigan's Upper Peninsula had a very high level of genetic similarity. They used their data to argue that these samples constitute parts of one gigantic fungus. Estimating that the biomass of the fungus, most of which was located underground and connected by rhizomorphs, was more than ten tons, and that the fungus was over 1,500 years old, they concluded their paper by saying that "members of the fungal kingdom should now be recognized as among the oldest and largest organisms on earth." This claim might be (indeed, has been) reasonably disputed on the grounds

47

that the fungus is neither a continuous structure nor an individual with determinate growth patterns. In order to enter this debate, let alone to resolve it, we need some criteria for answering the question "What is an organism?"[2]

A different sort of case concerns pregnant mammals. Pregnant females contain an embryo that becomes a fetus through gestation, with the fetus developing through the transfer of resources from mother to fetus via the placenta. Since this transfer of resources is to the physiological detriment of the mother, and the placenta is constructed from fetal, not maternal cells, the placenta is often viewed as a device of the fetus that wrestles control of internal resources from the mother. But the placenta is no mere device, for it is a living agent in its own right, not only signaling the needs of the fetus but actively intervening in the physiology of the mother, one side effect of which is sometimes diabetes, amongst other physiological complications, in the mother. In addition, when there is a shortage of resources, the placenta harnesses them for its own growth at the expense of the fetus, which subsequently becomes malnourished. Finally, many of the genes needed for the fetus to develop are those of the mother (maternal effect genes), and the mother and fetus constitute a physiologically integrated individual. The Australian geneticist David Haig's work on genetic imprinting – genes expressed only if they derive from one or the other of the parents – has introduced the idea that pregnancy can profitably be understood as a form of parent-offspring conflict, with the placenta as a mediating organ. Following his Harvard predecessor, Robert Trivers, Haig has modeled this conflict using the perspective of the selfish gene.

Answering the question "What is an organism?" is not strictly necessary to make progress on understanding many of the genetic, physiological, and evolutionary intricacies of the example. But that question is certainly raised by these intricacies. If we begin, naively, by thinking of the mother as a clear case of an organism, the fetus less so, and the placenta merely as a living agent, these details may provide a basis for revising our views here. But why? Issues about control, physical integrity, dependence, interest, and conflict are surely part of what drives and shifts intuitions here. Charting the relationship between these and the concepts of an organism and a living agent is part of the task of answering our eponymous question.[3]

I present the final example in more elaborate detail. Living coral reefs consist of two components. They are accretions of calcite deposits produced by polyps that, in turn, grow on this calcium-based foundation. The polyps are organisms, millions of which produce the organic deposits that constitute the calcified environment that they and their descendants literally live on. It is natural and common to describe coral reefs in this way, and not all that puzzling even to think of both the polyps and the complex they form, the living reef, as organisms.

Although polyps themselves are capable of producing calcium carbonate, it turns out that to produce it in the abundant quantities necessary to build a reef, the presence of protozoan flagellates known as *zooxanthellae* are necessary. These single-celled organisms infect the polyps in what is a symbiotic relationship, producing glucose that provides energy to power the process of calcification, and gaining in exchange a sheltered food den in which to make their own living. Zooxanthellae contain the pigments that give living corals their beautiful colors, and their absence is a clear sign that not all is well with the reef. So we now have at least two paradigmatic organisms, the zooxanthellae and the polyps, with the former living not on but literally *in* the latter. Perhaps these together with the calcite deposits constitute the living coral reef.

The real puzzles start when we begin to probe further into the energetics of each of these biological agents and how they manage to earn their keep. Zooxanthellae photosynthesize and they require carbon dioxide in order to do so. When carbon dioxide dissolves in water, it produces carbonic acid (H_2CO_3), which then dissociates into a hydrogen ion and bicarbonate (HCO_3^-). Polyps require carbonate (CO_3^{-2}) in order to produce calcium carbonate, which means that hydrogren ions need to be stripped from bicarbonate. The problem is twofold.

First, although the reaction from carbon dioxide through carbonic acid to bicarbonate is reversible, as is that from bicarbonate to carbonate, both reactions are biased to produce more bicarbonate than anything else. Hence, energy is needed to move the reaction process from bicarbonate, which is not itself used either by zooxanthallae in photosynthesis or by polyps in calcification. Second, the desirable outcomes of this reaction – to produce carbon dioxide (for zooxanthellae) or carbonate (for polyps) – stand at opposite ends of the overall reaction. Thus, the two organisms must drive the reaction in opposite directions. The solution to both problems is to shuffle hydrogen ions away from carbonate once it is formed (to the benefit of the polyps), and to bicarbonate to drive carbon dioxide production (to the benefit of zooxanthellae).

This process is not fully understood, but enough is understood to raise some questions and chalk out some positions. J. Scott Turner has argued that this represents an example in which "the boundary between

organism and NOT-organism is not so clear. An important component
of the physiological process takes place outside the animal" – the polyp.
Turner uses the example to illustrate his more general thesis that "the
edifices constructed by animals are properly external organs of physiology." I shall return to discuss Turner's broader views in the next chapter,
but here I want to note that the example itself raises questions whose
answers require some understanding of what organisms are that goes
beyond pointing to paradigm examples. Could the physical boundary of
the organism be irrelevant to where its physiology begins and ends? Is
there any sense in which the whole system – polyp, calcite, and zooxanthellae – constitutes a clearer example of an organism than do any of its
components? Where do organisms begin and end?[4]

2. ORGANISMS AND LIFE: THE SIMPLE VIEW

As a start on the conceptual spadework that needs to be done, I begin
with an intuitive view of organisms that takes seriously the central role that
organisms play in our thinking about living agents. As its name suggests,
it is not the final answer to the question "What is an organism?" But
discussion of its shortcomings will set us on the right track.

I call this view the *Simple View* of organisms. It says that organisms just
are living agents. Recall that in Chapter 1 I distinguished between living
and nonliving biological agents, including amongst the latter both the
very small, such as genes and proteins, that are typically parts of organisms, and the very large, such as demes and species, of which organisms
are a part. The Simple View holds that organisms are not simply paradigmatic but in fact exhaust the realm of living agents: only organisms are
living agents. Conversely, the view also implies that all organisms are living agents. What makes the Simple View prima facie plausible, and what I
think is right about it, is that it recognizes a close conceptual connection
or metaphysical relationship between organisms and living agency.

If the Simple View posits a kind of *a priori* conceptual analysis, it recognizes the former, and for philosophers who like their conceptual analyses
short and snappy, it is about as short and snappy as they come. How informative it is, however, turns on what is built into the concept of a living
agent, and there is a notorious trade-off between the snappiness of conceptual analyses and their informativeness (not to mention their truth). I
said in Chapter 1 that I thought of an agent as something like a physically
bounded locus of causation, and that living agents were typically characterized in terms of a range of properties, such as metabolism, growth,

and adaptation. It is conceivable, I suppose, that the Simple View could be developed in ways that build on these views in plausible and informative ways, as has been done with many ordinary concepts in the hands of analytic philosophers. But since this view of the task of philosophical analysis has not produced a single, widely accepted analysis in over one hundred years, it would seem like wishful thinking to see the Simple View as likely to provide a strictly true analysis of the concept organism.[5]

If, by contrast, the Simple View makes an *a posteriori* empirical identification, it recognizes a close metaphysical relationship, that of identity, between organisms and living agents. This is the kind of relationship that one might recognize (to take two well-worn examples) between lightning and electrical discharge, or between creatures with a heart and creatures with a kidney. Applying this to the Simple View, although the concepts "organism" and "living agent" are distinct concepts, they have precisely the same reference in that they pick out exactly the same things in the world. Because of this, it is correct to identify organisms with living agents, and doing so can be informative because we can apply what we know about the world subsumed under one concept to the other. Thus, we can develop a deeper, more integrated view of biological agents by adopting the Simple View.

Philosophers who view themselves as enlightened by the Quinean naturalistic turn in philosophy are sometimes scornful of the first of these sorts of view but happy enough to adopt the second. Since both general views are subject to many of the same problems and, as I shall argue shortly, face the same problems when applied to the Simple View, I take this perspective to be problematic but largely irrelevant to our discussion. The general problem with the Simple View, however it is construed, is that it overstates the relationship between organisms and living agents. In so doing, it fails to recognize the heterogeneity of both of these central biological categories.[6]

3. PROBLEMS FOR THE SIMPLE VIEW

The Simple View entails that all and only organisms are living agents. Both of these entailed claims are false. Consider first the "all" claim. According to it, strictly speaking there are no dead organisms. The obvious way to obviate this problem is to modify the Simple View so that it says that for at least at some point during their existence organisms are living agents. This is no doubt true, but note that it takes us beyond the Simple View. The same general point is true of the (different) changes required to account for insect colonies that are considered as superorganisms, if these are examples of organisms that are not themselves living.

Consider the claim that *only* organisms are living agents. There are at least two kinds of biological agents that are strong candidates for being living things but are not themselves organisms. The first of these are relatively large parts of organisms, such as organs or physiological systems. The second are organelles, such as mitochondria, which were once, in the evolutionary past, organisms, but are no longer. Intuitively, what both lack is the sort of independence, autonomy, or free living that organisms have, since their existence as living agents is not simply tied closely to that of other living agents, organisms, but dependent on it. We might modify the Simple View in order to account for these putative counterexamples, but again this suggests the need to move beyond that view.

The deeper problem facing the Simple View is that it fails to show a grasp of what I shall call the *intrinsic heterogeneity of the living world*. As the three examples in the first section indicate, there are difficult questions about what counts as an organism, about the role of physical boundaries in individuating organisms, and about where organisms begin and end that are raised by moving beyond the most obvious instances of organisms. I have also mentioned other examples of living agents, such as organs, physiological systems, and previously free-living organelles. Many of the same sorts of questions arise about living agents: what demarcates them from nonliving agents, how are they individuated, where do they begin and end? In addition to those agents already mentioned, the living world also contains obligate parasites, single-cell organisms, plants, asexually and sexually reproducing organisms, sterile insect castes, archaebacteria, and perhaps even some of the following: some kinds of virus, developmental systems, creatures devised in artificial life programs, and (under the Gaia hypothesis) the whole planet. Any adequate account of either organisms or living agents needs to provide some guidance about this diverse range of cases, and what to say about them. The Simple View, in identifying the two, makes little progress on this front.[7]

One response to the problem of intrinsic heterogeneity is pluralism about either organisms or living agents (or both). The categories of organism and living agent encompass distinct kinds of things, and deploying a more fine-grained scheme of individuation is necessary to do justice to the diversity we find. As elsewhere in the biological sciences, such a pluralistic move has been made by both biologists and philosophers, and can be more or less general in scope. For example, in the context of understanding vegetative growth, the botanist John Harper has distinguished between what he calls *ramets* – units of clonal growth – and *genets* – the reproductive output of either sexual or asexual reproduction. Harper uses the distinction as the basis for replacing questions about how many individuals exist and where their boundaries are, in given cases, with questions about ramets and genets. Likewise, the philosopher Jack Wilson has distinguished at least three different kinds of individuals subsumed under the heading "living kinds" – genetic, functional, and developmental. Wilson argues that our thinking about what I am calling living agents and organisms benefits from adopting his pluralistic framework in place of a misleadingly simplistic, monistic view of either.[8]

One final problem for the Simple View – one hinted at in the previous section and related to the previous problem concerns how informative the view is. We would like an analysis to carve up the world in ways that make sense, but we also need it to tell us what, if anything, is special about organisms. To add to the informativeness of the Simple View, we need to say much more about living agents. Agents themselves constitute a diverse assortment of entities, and the pluralistic views mentioned previously employ more fine-grained categories as a way to develop more informative views, but at the expense, I think, of telling us what is special about either organisms or living agents, despite the diversity amongst them. I indicated in Chapter 1 that there was no principled reason not to further extend the realm of agents from thinglike entities or individuals to processes, events, or even properties. But if we operate with the more restricted notion of an agent that I have invoked, then I think there is something special about biological agents as a class, living agents within that class, and organisms as types of living agents. All three are natural kinds, a point as simple sounding as is the Simple View, but which points to the way between that view and pluralism.

To head down this path, a path leading to a more adequate answer to the question "What is an organism?," I need to say something general about natural kinds and the realist views in which they have, for the most part, been discussed.

4. NATURAL KINDS, ESSENTIALISM, AND SCIENTIFIC REALISM

The idea that there are natural kinds has a history in and an aptness for articulating realist views of science. Realists have traditionally held something like the following view of natural kinds. Natural kinds are what the sciences strive to identify; they feature in laws of nature and so scientific explanation; they are individuated by essences, which may be constituted by unobservable (or "theoretical") properties; and they are conceiver-independent classifications of what there is in the world – they "carve nature at its joints."

The traditional realist view of natural kinds extends the following naive, common-sense view. There are objects and properties that exist independently of human observers. For example, suppose that we have before us a piece of rock. It has properties, such as having a certain mass and constitution, and the rock together with its properties exist independently of human observers. Scientists investigate such objects, uncover certain relationships between their properties, and develop taxonomies, natural kinds, that make these relationships more apparent. Suppose our rock has the property of being made of molten lava (composed, say, of 50% silica) and so has a certain melting point and various other chemical properties. By taxonomizing it as an *igneous* rock, scientists can both recognize its relationship to other kinds of rock and explore the relationships between the properties that igneous rocks have.

The traditional realist view of natural kinds goes beyond such a common-sense view chiefly in the depth of its metaphysical commitments. Distinctive is the realist's view of why certain relationships between properties hold, and why scientific taxonomies that identify natural kinds reveal further relationships between properties. Some properties are coinstantiated or correlate with one another because they feature in laws of nature, and these hold because of how nature is structured. In addition, the properties that feature in laws of nature are intrinsic properties of the entities that have them: they are properties that would be instantiated in those entities even were those entities the only things that existed in the world. This traditional realist view is, thus, individualistic about the properties that individuate or taxonomize natural kinds, essences (and is so in all three senses introduced in Chapter 1).

On this view, natural kinds categorize objects in terms of the intrinsic properties they have: same intrinsic properties, same kind of thing. And this explains why taxonomies that identify natural kinds lead to further revelations about how properties are related to one another, assuming that the most fundamental properties in the world are intrinsic properties. I want to focus on two further aspects to this overall metaphysical conception of natural kinds that I introduced in Part One in discussing recent views of species: essentialism and unificationism.

Essentialism is the view that natural kinds are individuated by essences, where the essence of a given natural kind is a set of intrinsic (perhaps unobservable) properties, each necessary and together sufficient for an entity's being a member of that kind. Realists thus say that scientific taxonomy proceeds by discovering the essences of the kinds of things that exist in the world, and that this explains, in part, the theoretical and practical successes of science. The endorsement of essentialism provides a way of distinguishing natural kinds from arbitrary and conventional groupings of objects. Natural kinds are *kinds* (versus mere arbitrary collections) because the entities so grouped share a set of intrinsic properties – an essence – and *natural* (versus conventional or *nominal*) because that essence exists independently of human cognition and purpose.

As a general thesis, unificationism is the view that scientific knowledge is unified in some way. For the traditional realist, it is the view that since natural kinds reflect preexisting order in the world, they are unified or integrated. But realists are not alone in holding some version of unificationism about scientific knowledge. The strongest versions of unificationism were held by the logical positivists as the "unity of science" thesis (for example, Oppenheim and Putnam 1958), and came with a reductive view of the nature of "higher-level" scientific categories. More recent unificationist views have been nonreductive, cast in terms of the notions of constitution or realization, rather than identity.[9]

Traditional realism, whether in its reductionist or nonreductionist guise, implies views about the basis of membership in a given natural kind, the relationship between the various natural kinds and the complexities in nature, and the ordering of natural kinds themselves. We might express these views as follows:

> *commonality assumption*: there is a common, single set of shared properties that form the basis for membership in any natural kind;
> *priority assumption*: the different natural kinds there are reflect the complexities one finds in nature, rather than our epistemic proclivities;
> *ordering assumption*: natural kinds are ordered so as to constitute a unity.

For a traditional realist, the commonality assumption leads naturally to essentialism about natural kinds. The priority assumption points to the world, rather than us, as the source of the variety of natural kinds one finds. And the ordering assumption, typically expressed in the view that natural kinds are hierarchically organized, says that there is one way in which different natural kinds are related to one another.

5. NATURAL KINDS AS HOMEOSTATIC PROPERTY CLUSTERS: LIVING AGENTS

One problem for the traditional view of natural kinds as having intrinsic, physical properties as essences is that, as we saw in Part One, many biological categories are relationally individuated. Thus, if the traditional view is to accommodate this aspect of the biological world and how it is conceptualized it will at least need to appeal to relational essences.

But a larger problem looms. If we take the intrinsic heterogeneity of the living world seriously, then even a form of essentialism liberated from its individualistic shackles will not suffice. For biological kinds generally will each subsume individuals that differ from one another, even with respect to putatively essential properties. I shall attempt to fill out and justify this claim with respect to the kind living agent, and to show how to modify traditional essentialism within an overarching realist framework with this problem in mind.

Recent work in naturalistic epistemology based on the philosopher Richard Boyd's claim that at least some natural kinds are *homeostatic property cluster* (HPC) kinds provides a ready-made way to conceptualize the natural kind living agent. The basic claim of the HPC view is that natural kind terms are often defined by a cluster of properties. No one or particular n-tuple of this cluster need be possessed by any individual to which the term applies, but some n-tuple of the cluster must be possessed by all such individuals. The properties mentioned in HPC definitions are *homeostatic* in that there are mechanisms and constraints that cause their systematic coinstantiation or clustering. Thus, an individual's possession of any one of them significantly increases the probability that that individual will also possess other properties that feature in the definition. This is a fact about the causal structure of the world: the instantiation of certain properties increases the chance that other particular properties will be coinstantiated because of underlying causal mechanisms and processes.[10]

The view is a "cluster" view twice over: only a cluster of the defining properties of the kind need be present for an individual to fall under the kind, and such defining properties themselves tend to cluster together, that is, tend to be coinstantiated in the world. The first of these features of the HPC view of natural kinds allows for inherent variation among entities that belong to a given natural kind. And the second of these features distinguishes the HPC view as a realist view of kinds from the Wittgensteinian view of concepts in general to which it is indebted. On the HPC view, our natural kind concepts are regulated by information about how the world is structured, not simply by conventions we have established or language games we play.

To see how to apply the HPC view to living agency, take the sort of properties that I have already mentioned as characterizing living agents to constitute the relevant cluster of properties that define the kind. On this view, living agents are defined by something like the following cluster of properties. They are causally integrated entities with a physical boundary that are a locus of causation (agents) and:

- have parts that are heterogeneous and specialized
- include a variety of internal mechanisms
- contain diverse organic molecules, including nucleic acids and proteins
- grow and develop
- reproduce
- repair themselves when damaged
- have a metabolism
- bear environmental adaptations
- construct the niches that they occupy

This set of properties forms a homeostatic cluster in that there are mechanisms and constraints that promote the coinstantiation of many of them. Even were some of these properties (for example, having heterogeneous parts, or a metabolism) universally shared by all living things, it is the cluster of properties that informatively defines the kind, not any member of that cluster. While it is unlikely to be the ultimate list of properties in the cluster for living agent, it would surprise me if that ultimate list departed radically from the one above.

Next, we identify individual things as members of the kind living agent in terms of their possessing some sufficient subset of the properties in that cluster. Different living agents may share different subsets of the cluster. Sterile organisms do not themselves reproduce. Some single-celled organisms grow and develop minimally or only during a restricted stage of their existence. And many plants radically restrict their metabolism during seasonal variation. There will be an inherently vague boundary in membership in the kind because of the vagueness in the idea of a sufficient subset of properties.

Consider physical things often regarded as borderline cases of living agents, such as viruses, many of which are structurally little more than stretches of nucleic acid with a protein coating, and "self-replicating" protein chains, which lack organelles and an internal physiology. Their borderline status is readily explained by the HPC view: they have some of the structural and functional properties in the cluster that characterizes living agents, but lack many others. The HPC view of living agents provides a natural way of accounting for such borderline cases. But, as importantly, it also implies that the demand for clean, crisp categories that arbitrate on such cases is a mistaken philosophical ideal imposed on a complicated biological world.

This realist view of natural kinds has been developed primarily with respect to species, and I shall return to it again in discussing species in Chapter 5. But it should be clear that it is particularly apt for characterizing inherently heterogeneous kinds of thing whose individual variability may be critical to their being the kinds of things that they are. For such entities, taxonomy cannot proceed with a set of individually necessary and jointly sufficient properties. And since the biological sciences are largely concerned with identifying both the underlying causal mechanisms and shared constraints that govern biological processes and the properties they involve, the HPC view comports with the explanatory focus of the biological sciences. It thus provides a relaxation of traditional essentialism and realism about kinds in a way that acknowledges intrinsic biological heterogeneity.[11]

While the properties in the HPC kind living agent that I have listed previously are intended to be indicative rather than definitive, it may pay to reflect a little more on the types of properties that appear on that list. They can be categorized into three familiar types in a way that highlights the homeostatic relationships between them.

First, the first three included in my characterization of the HPC kind – having heterogeneous, specialized parts, a variety of internal mechanisms, and containing organic molecules – are *structural* properties. As complex, causally integrated physical objects, living agents have a structure, one that partially determines what they can and cannot do.

Second, properties such as growth, reproduction, and repair are *functional* properties that concern how these internal structures operate, what they do. These account for what we might think of as the internal housekeeping that living agents perform, the physiology of the living agent. They can be thought of as dispositions that living agents have, capacities that they are able to realize once they are put into specific environmental contexts, in virtue of the structural properties they possess.

Third, the final properties that I listed – metabolism, environmental adaptation, niche construction – concern the interaction of a living agent with its environment, and presuppose the existence of both specific structural and functional properties. These properties are relational and, drawing on the terminology of Chapter 2, they have wide realizations. They are instantiated in systems that physically extend beyond the boundary of the individual organism that possesses or bears them. Since they concern how a living thing functions, I shall call them *wide functional* properties. They need to be understood through integrative synthesis, rather than constitutive decomposition.

6. THE TRIPARTITE VIEW OF ORGANISMS

Both the Simple View and traditional essentialism about natural kinds contain something that is correct, and much of the chapter thus far has been directed at building on this. It is now time to state and then defend my own answer to our initial question, "What is an organism?"

I call this the *Tripartite View*. Baldly stated, the Tripartite View says that an organism is:

    a. a living agent
    b. that belongs to a reproductive lineage, some of whose members have the potential to possess an intergenerational life cycle, and
    c. which has minimal functional autonomy.

I have already said enough, I hope, about a. Consider, in turn, b and c. To give them slightly more meaningful names, I shall refer to them, respectively, as *Life Cycles* and *Autonomy*.

7. LIFE CYCLES

Organisms are living agents, but not the only living agents. I have already mentioned various parts of organisms, and to the examples of these already provided we could add external growths such as leaves or body hair and skin. We often speak of these things as living (or as having died, and so as having been alive), but do not view them as organisms in their own right. That, I want to suggest, is because they fail to satisfy Life Cycles. (They may also fail to satisfy Autonomy, a point to which I shall return.)

Bodily systems and bodily products do not, like organisms, themselves belong to a reproductive lineage of living agents, some of which have the capacity to instantiate a life cycle. While such systems and products are replicated when organisms reproduce, they do not reproduce members of a lineage. They are replicated but are not themselves the agents of replication, not what reproduces members of the lineage. To see this more clearly, focus on a given token bodily part of an organism – say, a heart. That heart is not part of a lineage, some of whose members are living and reproduce. Any given heart reproduced in organisms forms part of a lineage of hearts, but none of the members of that lineage are themselves reproducers of hearts. In short, what organisms have that their bodily parts and products lack is the potential to possess an intergenerational life cycle.[12]

A distinction that the philosopher James Griesemer introduces in discussing genetics and development may be useful in explaining this point. Griesemer distinguishes between the processes of replication, a form of copying, and reproduction, which involves something more. That something more is twofold: reproduction involves both *progeneration* and *development.* Progeneration involves the multiplication of like entities through a process of material overlap, while development is just the capacity to acquire the capacity to reproduce. As Griesemer argues, biological reproduction through fission and fusion involves progeneration and not mere replication. Thus, the notion of a reproducer, rather than that of a replicator, is central to understanding how evolution works. In Griesemer's terms, my point is that bodily organs and products, such as hearts, may well count as replicators but they are not reproducers, and it is this latter notion that is connected directly to that of a life cycle.[13]

Intuitively, a life cycle is a series of events or stages through which a living agent, particularly an organism, passes. But which events or stages? Not all events in the life of an organism comprise its life cycle, but only those that are reliably replicated across generations. These replicated events are temporally bounded by one and the same kind of event, an origination-completion event, such as the formation of a fertilized egg in sexually reproducing organisms, or the creation of a fissioned cell in clonally reproducing organisms. Put more carefully, a life cycle is comprised of a causal succession of entities, each a living agent, which themselves, together with the processes that mediate their succession, recur across generations. "Development" names these mediating processes, and it is important to characterize life cycles generally enough to include the variety of forms that development takes, including somatic embryogenesis and more familiar epigenetic forms of development. An organism is the paradigmatic entity that has or possesses a life cycle, and the entities that comprise this cycle are stages in the life of that organism.[14]

It has been traditional to think of the processes that govern such cyclical replication as exclusively physiological and genetic. But there seems no reason to preclude other, external processes, including those that are cultural, social, or ecological in nature, from playing this mediating role, a point I shall discuss further in Part Three.

So an organism possesses a life cycle, and is at least a partial realization of what physically constitutes the life cycle it possesses. The question is whether this is also true of any parts of an organism, including its DNA or other developmental resources it deploys, that we would not normally think of as themselves organisms. If it is, then this would count against the Tripartite View.

Relatively short strands of nuclear DNA form lineages in that there is a series of processes – transcription, expression, polymerization, meiosis, translation – through which they are reliably replicated. Moreover, these replicative processes are cyclical. The same is true of the reproduction of other developmental resources, such as extranuclear organelles, the chromatin marking system, and even environmental and social structures, such as nests and parenting practices that are socially transmitted. Unlike organisms, however, these entities do not possess life cycles, either because the entities that occupy the stages in their replicative cycles are not living agents (for example, DNA) or because they do not progenerate and develop (for example, mitochondria). They are, like organisms themselves at any particular developmental stage, partial realizations of a life cycle, but this is the life cycle of the organism that they physically constitute, not their own life cycle.

Life Cycles provides the basis for understanding the emergent endosymbiotic status of some organelles, such as mitochondria and chloroplasts. These organelles began their existence as free-living organisms, becoming incorporated into eukaryotic cells over one billion years ago. No doubt gradually, they gave up their own roles as reproducers, making use instead of the more powerful reproductive apparatus of the new organisms that they partly physically constituted. If the Tripartite View is correct, then we can see their transition from organism to organelle as a function of their shift from being living agents of which Life Cycles is true to being living agents of which it is false.[15]

The idea that the notion of a life cycle plays a crucial role in characterizing what an organism is has been recognized by others. But this role has sometimes been mischaracterized even by those most familiar with it. John Tyler Bonner has done more than anyone to place the concept of the life cycle at center stage in thinking about organisms and their development. Bonner goes so far as to identify organisms with life cycles, as do the developmental systems theorists Paul Griffiths and Russell Gray. Yet if we think of reproduction as a transitional process from one token life cycle to the next, as is common, then since many organisms continue their life after reproduction, or even without it, it is difficult to see how one can strictly identify organisms with life cycles. Organisms are not identical with life cycles, I have suggested, but both possess and (partially) realize them.[16]

Likewise, we should be cautious in how the life cycle of organisms is described. Bonner's characterization of "the life cycle" as having four temporal, successive periods – the single-cell stage, growth and development, maturity, and reproduction – captures something true of much multicellular life. Yet single cell organisms live their whole lives as single cells. Many organisms grow after maturity, and many do not reproduce at all. Also, as Bonner himself notes in passing, reproduction is not necessarily contained in a period, particularly not one that follows maturity. Furthermore, many organisms do not have even the capacity to reproduce (for example, sterile castes of "social insects"), though some do have the capacity to acquire this capacity under the right circumstances. If such sterile organisms possess a life cycle, then either reproduction is not part of that cycle, or their reproduction is mediated through the replicative activities of conspecifics (for example, female reproductives).[17]

Life Cycles posits a less direct relationship between individual organisms and the possession of an intergenerational life cycle, one that allows for nonreproducing or even sterile organisms. A full, intergenerational life cycle must be manifest within any given organism's lineage, but such a life cycle need not be completed within that organism's own life history. This may make some of the facts relevant to determining a given living agent's organismic status facts about other members of that agent's lineage (for example, its conspecifics), rather than facts about that agent itself. I take this to be another way in which biological categories violate the putative constraint of individualism in biology, and that suggests the plausibility of an externalist view of biological taxonomy.

8. AUTONOMY

Even were possession of a full, reproducible life cycle strictly necessary for any living agent to be an organism, there would remain more to being an organism than that. Organisms are not simply living agents that can reproduce, or that form parts of reproductive lineages. For in addition, a living agent must have what I am calling minimal functional autonomy.

Although I introduce the concept of minimal functional autonomy as a technical notion, it is intended to build on an intuitive notion of autonomy that operates in our thinking about living agents. It involves two complementary components. Each of these acknowledges the bodily nature of living agents that are organisms, and builds on the idea that all living agents are physically bounded loci of action. Some of these, organisms, share a level of autonomy that the merely living lack. Organisms are to some extent free from what lies beyond their boundary, and have control over that which lies within that boundary. In common terms, organisms *have a life of their own*; they exercise control over themselves and thus are at least to some extent free of both the agency of others and the action of the world more generally.

Consider some examples of living things that are not organisms to see what Autonomy is saying. Organs and bodily systems are living, but they are not minimally functionally autonomous of the individuals they comprise. As I have said, they have a life that in the ordinary course of things is dependent on that of the agents they are housed in. The possibility of organ and tissue transplants show that there is some flexibility in this dependency relationship, though the restrictive conditions in which such transplants are successful suggest just how constrained the life of organs and tissues is by that of the organisms they constitute. In this respect, cells might be thought to have more autonomy and control, since they can be cultured outside of any organism. Yet cultured cells typically lose much of their functionality and sometimes the control over their own internal processes, once they are removed from their organismic environment.

Much the same might be said of obligate parasites or symbionts, who have, in the extreme, ceased to be the locus of control for their own activities, much in the way that they have abandoned having their own life cycles. Their tight dependence on other organisms sustains whatever kind of life they have. To the extent that they do maintain autonomy and control, they remain organisms.

Consider cases of endosymbiosis. Bacteria within the *Buchnera aphidola* clade form part of the digestive system of their aphid hosts, who in turn transmit the bacteria to offspring cytoplasmically. The bacteria then form part of the developing digestive system of those offspring. *Buchnera* supply amino acids that are missing from the natural diet of the plant saps that aphids consume, and aphids have demonstrated control over the distribution of *Buchnera* amongst their eggs. When *Buchnera* are not transmitted to offspring, such as when the aphids are treated with antibiotics that kill the bacteria, the offspring suffer because of their compromised digestive systems. The association between *Buchnera* and aphids is roughly 250 million years old, and it has likely allowed aphids to exploit niches that would otherwise lie beyond their nutritional reach, and so facilitated speciation. Yet unlike the case of the formerly free-living bacteria that became mitochondria, the *Buchnera* clade shows only minimal genetic reorganization from related free-living bacteria, such as *E.coli*.[18]

There are two features of Autonomy that both help to further explain what minimal functional autonomy amounts to, and (for better or worse) that complicate the relationships between organism, agency, and life. I shall simply identify these features here, leaving the complications for more detailed discussion in the next chapter.

First, autonomy and control are both complex matters of degree whose magnitude depends on what one chooses to focus on. A living agent is more or less autonomous from its external environment, and more autonomous in some respects than in others. Yet there appears to be no one scale on which these aspects to autonomy can be compared. They are incommensurable.

For example, an organism may be autonomous with respect to respiration or metabolism, but not so with respect to sex determination, which may be under the control of some other individual (such as an egg-laying female in *Termita*) or an environmental variable (such as temperature in *Crocodilia*). The behavior of water-dwelling *Paramecia* bacteria is autonomous of the effects of gravity because of the ratio of their surface area to their mass and how this interacts with the surface tension of water. But since *Paramecia* contain magnetosomes, their behavior is not autonomous of the Earth's magnetic field.

Second, each determinate instance of external autonomy and internal control is irreducibly *normative*. The level of autonomy and control that agents are viewed as possessing turns on the standards that we adopt in assessing their actions. These standards are often implicit, rather than articulated explicitly, but are there nonetheless. They are pervasive in ascriptions of Autonomy. Again, consider some examples.

Some organisms that can regulate their own temperature are autonomous with respect to thermoregulation. We call them "warm-blooded animals." But this is not an absolute property that organisms either have or lack. Rather, thermoregular control is a matter of degree, and there is an implicit standard against which ascriptions of the property are made. Cell division is never completely controlled by the organism, but we ascribe such control relative to cases, such as cancerous growth, in which it is lacking to a much greater degree. Strictly speaking, no organisms are free from the force of gravity. But as the example of *Paramecia* indicates, the ascription of autonomy here is naturally made because relative to the effects that gravity has on other organisms, those on *Paramecia* are negligible. Again, we have an implicit standard that allows us to make a categorical distinction when the reality described may be continuous.

Having introduced the Tripartite View and explained the respects in which it goes beyond the Simple View of organisms, I shall conclude this chapter with a brief discussion of how aspects of the Tripartite View are prefigured in the work of Herbert Spencer and Julian Huxley. Doing so will return us to some of the broader issues introduced in Part One, and direct us to several controversial implications of the Tripartite View.

9. TWO PRECURSORS: HERBERT SPENCER AND JULIAN HUXLEY

That Herbert Spencer and Julian Huxley had much to say about what organisms are will not surprise anyone even passingly familiar with their more general views, or the broader intellectual context in which each of them wrote. And those more familiar with their views who have read this far will already have noticed that my own views build on those of Spencer and Huxley. In part as intellectual acknowledgment but also in part to direct those working in the fragile sciences back to authors worth revisiting more generally, I make explicit here some of the connections between the Tripartite View and those of Spencer and Huxley.

Spencer discusses the question of what organisms are in the short, sixth chapter of the first volume of *Principles of Biology.* Spencer considers the limitation of contemporary answers to the question in correctly pronouncing on the individuality of the range of plants and animals including strawberry plants, the colonial flagellate *Volvox globator,* a variety of polyps and ascidians, and the Canadian waterweed *Eloidea canadensis.* Spencer warns that "[t]here is, indeed, as already implied, no definition of individuality that is unobjectionable" and thus that one should "make the best practicable compromise." He then offers his own view:

The distinction between individual in its biological sense, and individual in its more general sense, must consist in the manifestation of Life, properly so called. Life we have seen to be, 'the definite combination of heterogeneous changes, both simultaneous and successive in correspondence with external co-existences and sequences.' Hence, a biological individual is any concrete whole having a structure which enables it, when placed in appropriate conditions, to continuously adjust its internal relations to external relations, so as to maintain the equilibrium of its functions.[19]

Note two parallels with the Tripartite View.

First, for Spencer, 'individual in its biological sense' is to be understood in terms of life, just as I have suggested that we understand organisms as a kind of living agent. Second, Spencer has a principled reason (one that appeals to the nature of evolution) for expecting many gray areas where the biological grades into the nonbiological, and so he is not perturbed by the failure of any theory, including his own, to account for every example that one can produce. It is for this reason that he is happy to "make the best practicable compromise." I have suggested that the HPC view of living agents should lead us to expect (and indeed embrace) irresolvable, borderline cases of living agents.

Consider now Spencer's characterization of life, which he quotes above from an earlier chapter of the *Principles*. In an early paper, Spencer had characterized life simply as "the coordination of actions," and the idea of acting as a cohesive unit, an integrated whole, remained at the core of his later conception of life. Two ideas were, however, missing from this earlier view (or at best, implicit in it): that living agents coordinate heterogeneous changes within themselves, and that they coordinate these changes not only with one another but with the conditions of the external world. Let us take each in turn.[20]

Spencer suggests that the changes that occur within the biological world are more diverse, more varied from one another, than those that occur within the nonbiological world. Moreover, even when, in particular cases, this is not true, biological processes are combined with one another in a more highly coordinated manner. Despite their heterogeneity, they function as a unit. As Spencer says, "[R]espiration, circulation, absorption, secretion, in their many sub-divisions, are bound up together.... But we miss this union among non-vital activities." This sort of functional integrity unifying structurally and functionally diverse parts makes embodied autonomy possible.[21]

Spencer's correspondence condition aims to capture the responsiveness of living agents to environments beyond their own boundaries, to things distinct from themselves but with which they are in an ongoing, reciprocal causal relationship. It is to highlight this externalist aspect to his view of life that Spencer restates his view, in succinct form, as characterizing life as "[t]he continuous adjustment of internal relations to external relations." In the terms introduced here, this adjustment is the result of the wide functional properties that characterize living agents.[22]

As I mentioned in Part One, the phenomena of life and of mind were taken by Spencer to be inextricably interwoven. Spencer views both as "kinds of vitality," the two kinds that are "most unlike" bodily vitality and intelligence, or what we might call *the life of the body* and *the life of the mind*. One might think that, particularly so expressed, Spencer's view here presupposes the bifurcation between material body and nonmaterial mind, but this would be a misleading impression. For as Spencer's early commitments to an evolutionary view of both the biology and psychology of organisms makes clear – for example, as manifest in Parts I and II to Volume 1 of his *Principles of Psychology* – Spencer views both life and mind in "higher" organisms to have developed from those of "lower" organisms. His two kinds of vitality are equally realized in a material substrate, despite his own reservations about the "substance of the mind."[23]

I have already noted that we no longer view either life and mind or biology and psychology as intimately connected in this way. In particular, biology is typically conducted independently of psychology, and living agents are mostly things without minds. Thus, we may simply take Spencer's views here to be a quaint artifact of the grand, Victorian context in which he wrote, akin to his attribution of various degrees of vitality to organisms in the Great Chain of Being. However, although this bifurcation of the biological and the psychological is part of official disciplinary rhetoric, as I said in Chapter 2, psychological language is pervasive throughout biology. As we will see in the next chapter, living agents are often described as if they had minds. There is at least the remnant of the sort of connection that Spencer sees between the two in contemporary biology.

Julian Huxley's *The Individual in the Animal Kingdom*, develops a number of Spencerian themes but goes beyond Spencer in several important ways. Huxley suggests three conditions of minimal biological individuality:

the individual must have heterogeneous parts, whose function only gains full significance when considered in relation to the whole; it must have some independence of the forces of inorganic nature; and it must work, and work after such a fashion that it, or a new individual formed from part of its substance, continues able to work in a similar way.[24]

Like Spencer, Huxley here appeals to heterogeneity as a mark of the biological, but for distinctive reasons. In contrast to Spencer's postulation of the heterogeneity of the biological as a sort of inductive generalization from observations of the biological and nonbiological worlds, Huxley argues that biological individuals, organisms, must have heterogeneous parts; he does so by linking heterogeneity and the unity of the organism together. It is this unity that distinguishes organisms both from their parts (biological nonorganisms or nonindividuals) and from individuals that are not part of the living world (nonbiological individuals). Huxley's thought here comports with the ideas about the relationship between organisms and agency that I have posited, and I want to probe it a little here.[25]

Material objects, such as mountains and solar systems (two of Huxley's examples) are aggregates of similar parts, such that when one removes a part of the whole, the remainder continues to be the same sort of thing that it was, functioning similarly and differing only in degree from its existence before. This is the sense in which such natural objects constitute mere aggregates, rather than unities. By contrast, organisms constitute wholes, and their holistic nature entails and is entailed by the heterogeneous character of their parts. Without diverse parts, physical things are mere aggregates, not unities, and given that diverse parts have distinct but interlocking functions, they cannot exist and so function without forming part of a more unified whole, an organism.[26]

Huxley's contrast between organisms, their parts, and nonliving physical objects does not hold in general. For example, artifacts constitute one large class of counterexamples. However, Huxley is right to emphasize the structural (and hence functional) heterogeneity of the parts of organisms as something that makes attributions of Autonomy to organisms almost irresistible. For heterogeneity of parts does underlie the unity of the whole, something shared by organisms and artifacts.

Finally, consider Huxley's third criterion in the above-quoted passage. Here Huxley links organismic function directly to not only the self-maintenance and continuity of the organism, as had Spencer, but also to a potential line of individual descendants. This condition is Huxley's basis for thinking that there are higher orders of individuality, such as one might find in an ant colony or in a whole species. Indeed, he calls "species-individuality" *individuality in time*, and organismal individuality *simultaneous* or *spatial* individuality, saying "that wherever a recurring cycle exists (and that is in every form of life) there must be a kind of individuality consisting of diverse but mutually helpful parts succeeding each other in time."[27]

We can summarize these views in the terms introduced earlier in the chapter. Spencer and Huxley share the view that living agents are structurally heterogeneous and functionally self-maintaining. In addition, Spencer shows an awareness of the role of wide functional properties in his view of life. Huxley recognizes the importance of an intergenerational life cycle in characterizing organisms, and so accepts something like Life Cycles. Spencer's emphasis on the interdependence of the properties that constitute life parallels the view of living agency as a HPC natural kind, and Huxley's contrast between individuals and their parts can be readily understood in terms of Autonomy.

Finally, Spencer's outdated linking of life and mind suggests a role for the cognitive metaphor in biology. I shall discuss this metaphor in the next chapter by exploring some of the complications to the Tripartite View of organisms.

4

Exploring the Tripartite View

1. REALISM AND AGENCY

Building on the insights of the Simple View of organisms, I have given the concept of living agency a central place in the view of organisms sketched in the previous chapter. That concept also plays a structuring role in *Genes and the Agents of Life* as a whole. It is now time to raise, and at least begin to address, a difficult issue about biological agency relevant both to the Tripartite View of organisms and to the book more generally. To put it plainly: is there an objective, mind-independent fact of the matter about agency? Or is agency, and just what counts as an agent – and so a biological agent, a living agent, an organism – in the eye of the beholder? The issue is whether a realist view of agents, a sort of default view for most biologists and many philosophers of science, can be defended.

Thus far, I have taken for granted a realist view of agency. This was true of Part One in adopting a working characterization of an agent as a physically bounded locus of action, together with appeals to kinds of agents, to paradigms of these different kinds, and to the idea that agents are organized hierarchically. The very question "What are the agents of life?" might be thought to presuppose that agents are mind-independent, objective features of the world, rather than figments of our imagination (like pink elephants) or products of our collective social efforts (like conventions or human social practices). Likewise, the starting point of Part Two has been the view that living agents are part of the furniture of the world, even in the cases that reflect some of the complexities of that world. The question "What is an organism?" deserves to be taken seriously in part because answering it seems necessary in determining, for example, whether the humungous fungus is an organism, why a fetus but not a placenta is an organism, and what might make one think that the complex of polyps, calcite deposits, and zooxanthallae together constitute an organism.

What is the alternative to a realist view of agency? A graphic way to characterize such antirealist views of agency is to say, adapting Daniel C. Dennett's talk of the intentional stance, that they make being an agent a function of whether we decide to adopt the *agentive stance* in particular cases. Whether something is an agent, or a particular kind of agent, is up to us, something that we make up or construct, rather than something we discover about the fabric of the world.[1]

On an antirealist view, agency might be thought of as something that we project onto the world, in much the way that some people project the property of disgustingness to (say) fungi growing between your toes or maggots crawling through a ripe piece of fruit. The philosopher John Mackie was well known for developing this sort of Humean view of moral properties, including an analogy to this very example. Moral properties appear to be objective but are really, according to Mackie, projections of our own subjective values onto the world. The question we shall pursue in this section is whether the property of being an agent, or being a certain kind of agent, is actually like this.[2]

Given the previous chapter's defense of the HPC view of natural kinds as part of a modification of traditional realism, and the identification of living agents as such a natural kind, I am committed to a realist view of living agents. What of organisms, as a kind of living agent? According to the Tripartite View, organisms are living agents that satisfy two further conditions: Life Cycles and Autonomy.

Life Cycles, recall, is not simply the view that some living agents have life cycles, but that they are members of reproducing lineages some of whose members – themselves or others – possess and realize a complete intergenerational life cycle. Determining whether this is true of any given living agent may be difficult. In some cases, it may not even be possible because there is missing – even irrecoverable – information about either that living thing or its lineage members. But there is every reason to think that there is an objective, stance-independent fact of the matter about whether any given living thing satisfies Life Cycles. Our limitations here are epistemic, rather than indicative of some deeper metaphysical gap in the world that is filled by our subjective projections.

Autonomy says that organisms have minimal functional autonomy, where this is understood in terms of internal control and external freedom. Unlike being a living agent and satisfying Life Cycles, whether something can objectively have Autonomy attributed to it can be reasonably challenged by antirealists. This is in part because of the two features of minimal functional autonomy that I sketched near the end of the previous chapter: the incommensurability of the respects in which something might have minimal functional autonomy, and its normativity. Let me explain each in turn, and the problem they raise.

There are many different bases for the attribution of minimal functional autonomy to organisms. Living agents might have minimal functional autonomy with respect to a variety of properties, including temperature regulation, or sex determination, or movement in a gravitational field. That itself is no grounds for doubting the realist view that there is an objective fact of the matter about whether a given living agent satisfies Autonomy. But three further features of the situation do provide such grounds. First, there is no common currency in which these distinct properties can be compared. Second, organisms always have minimal functional autonomy with respect to some of these properties and not others. And third, there is no objective way to prioritize or order these properties, such that some can be viewed as basic conditions for minimal functional autonomy and others as derivative or optional in some way. If these claims are true, as I have suggested they are, then our ascriptions of minimal functional autonomy are incommensurable, and there is nothing in the biological world itself that determines whether a living agent satisfies Autonomy.

Consider now the normativity of ascriptions of minimal functional autonomy. Control of one's insides and freedom from what lies outside one's boundaries are never complete or absolute. Our categorical judgments about whether (or when) a living agent has sufficient internal control and external autonomy depends on explicit or implicit standards or norms that we accept. We need not be antirealists about norms in general to hold that the norms and standards we invoke in making judgments of minimal functional autonomy are not a part of nature, but our imposition on nature in making sense of how it operates.

It is no coincidence that issues about minimal functional agency often lie at the heart of debates over the organismal status of some living thing. Consider the three cases with which I began the previous chapter: the humungous fungus, the mother-fetus-placenta triad, and living coral reefs. A large part of what drives intuitions in such cases, as evidenced in the reasons that people produce for their views, are appeals to something like Autonomy.

In the case of the humungous fungus, the central issue is whether the fungus is an organism or whether instead it is a population of individual organisms whose relationship to one another creates a merely apparent humungous (and long-lived) organism. This is largely a question of minimal functional autonomy, and the various, determinate forms it takes: genetic, reproductive, physiological, and ecological. On an untutored view of the mother-fetus-placenta triad, mothers are thought to satisfy Autonomy most clearly, the fetus less so (and more so as it develops), and the placenta barely at all. But these judgments stand in need of revision once the biology of the relationships between the three are probed, which reduces these apparent differences between the three. As in the case of the humungous fungus, in that of the coral reef considerations of Autonomy may provide a basis for thinking of the entire complex of polyp-zooxanthellae-calcite as a better example of an organism than its individual parts. But it is also clear that Autonomy pulls one the other way, since the constituent agents in both cases satisfy Autonomy in at least some respects. Were we to accept an antirealist view of Autonomy, this would imply that in at least these cases there is no objective fact of the matter about where to draw the boundaries around the organisms.

The incommensurability and normativity of ascriptions of minimal functional autonomy should give us pause, but it should be a measured pause. They rock our realism, but do not, I think, topple it. The most plausible realist response is to challenge both incommensurability and normativity as a basis for rejecting realism. The claim of incommensurability itself and the inference from it to antirealism about minimal functional autonomy rest on too schematic a view of Autonomy, and too demanding a view of what needs to be true if realism is to be defensible.

Consider properties such as *being highly evolved*, or the relation *having more complexity than*. These share much with the property of having minimal functional autonomy, including those that serve as the basis for adopting an antirealist view of it. There are diverse ways to measure these properties. For example, in the case of being highly evolved and its relational derivative, being more highly evolved than, we have measures such as internal complexity, number of base-pair changes, number of speciation events, and adaptedness to past and current environments. There is no single scale on which all such measures fit. And organisms or structures that rank highly on some measures need not rank similarly on others. Still, one might say, we can fix on some of these measures and make objective, pair-wise comparisons of the degree of evolvedness between structures, organisms, higher-taxa, and other biological agents.

As with the case of size, for evolvedness and complexity there may be no single scale on which all things can be compared, and no basis for thinking that any one scale (for example, a two-dimensional versus a three-dimensional scale) is preferable in all cases. Yet realists need only more local, contextualized comparisons to be objective, something that holds in all three cases.

There is a second tack that a realist can adopt also suggested by these analogies. The chief argument for an antirealist view of minimal functional autonomy appeals to a suite of putative features of that property as it is instantiated by living agents. A natural view for a realist who accepts the Tripartite View of organisms is to take these features to be a function of our limited knowledge. Much as our judgments of evolvedness, complexity, or size may appear to be arbitrary, conventional, or incommensurable until they are sharpened by the development of more determinate forms of each of the corresponding concepts, so too with minimal functional autonomy. If the Tripartite View is correct, then Autonomy points to a property distinguishing organisms from mere living agents. But that is a property that needs further articulation, and whose full complexity is yet to be revealed through biological investigation. Final judgments about minimal functional autonomy should turn on the details of such exploration.

An ideal outcome for a realist would be for minimal functional autonomy to turn out to be itself the basis for a HPC natural kind, like living thing, where there is a mechanism and constraint-driven clustering of the various bases for the kind. This would leave borderline and irresolvable cases, but nonetheless would provide an objective basis for ascriptions of Autonomy.

2. A ROLE FOR THE COGNITIVE METAPHOR

I ended Part One by introducing the idea of the cognitive metaphor, the metaphorical ascription of psychological states to agents that do not literally have them. Organisms are living agents: they behave, they act, they do things in the world. Some living agents, such as human beings, have psychological states, and some of the things that they do are caused by those states. Folk psychology is the best-known framework for explaining the actions of agents with psychological states, and extensions of folk psychology are ubiquitous in the cognitive sciences. Not all agents literally have psychological states, even though their actions are often explained by invoking a framework such as folk psychology. The question I shall address in this section is why this is the case. The answer, I shall suggest, concerns the role of the cognitive metaphor in shoring up our views of agency and minimal functional autonomy.

The cognitive metaphor is operative whenever psychological terms are used to describe actions or behaviors of nonpsychological agents, or to explain actions or behaviors not caused by psychological states. When entomologists describe body types in a species of ant as strategies that members of that species adopt, or explain individual workers as sacrificing their reproductive interests in favor of those of the queen; when botanists explain increased leaf area in a given species as aiming to maximize photosynthesis, or a given flowering plant as preferring the shade; or when microbiologists conceptualize organisms as recognizing the presence of a pathogen, or a virus as choosing hosts with diminished immune systems, they invoke the cognitive metaphor. Behavioral ecologists and sociobiologists regularly apply the cognitive metaphor in explaining the behavior of minded and nonminded creatures, including ourselves, and it is manifest in talk of cell recognition, neural memories, molecular signaling, preferential developmental pathways, the goal of maximizing genetic replication, and of biochemical systems as seeking equilibria.

The cognitive metaphor is ubiquitous in the life sciences. It is the significance of this fact that needs to be determined. Part of the perceived significance of the cognitive metaphor in biology turns on whether one thinks that it can be eliminated, and whether it is merely shorthand for noncognitive descriptions and explanations. But even supposing that the cognitive metaphor is shorthand or eliminable (in practice or in principle), there would remain the issue of *why* the metaphor is ubiquitous. What does this currency buy biologists, given that they have other economic options? My focus will be on this neglected question.[3]

The shift made in the previous paragraph – from the question of why the cognitive metaphor is ubiquitous to what benefits it bestows on biologists who make use of it in one guise or another – betrays a sort of Darwinian functionalism on my part. The cognitive metaphor is there because it serves, or has served, some function within the practice of biology. Thus, we should examine the effect that the cognitive metaphor has in order to answer the second, and so the first, of these questions.

What the cognitive metaphor does, to begin with a metaphor, is to *magnify biological agency*. That is, it provides a way of understanding what biological agents do, and how they act, that increases and amplifies their roles as agents. In adopting the cognitive metaphor, we are able to envisage biological agents as independent actors, as relatively self-contained and self-directed individuals that adjust their relationship to their environments in ways that aim to further their own goals. To put this without using the cognitive metaphor itself, this is to say that the cognitive metaphor creates or facilitates the impression of biological agents as acting in ways that are autonomous of both other agents and their own environments.

We can express this in terms of the Tripartite View of organisms and my discussion of it thus far. What the cognitive metaphor does, in effect, is to crystallize minimal functional autonomy. Attributing psychological states to biological agents that do not have them allows those biological agents to appear to satisfy Autonomy. That is, it shifts them from the realm of merely living things to full-blown organisms. I suspect that this shift depends on the paradigm status that human agents have as both organisms and psychological agents. We literally have psychological states and by treating nonminded biological agents as if they had such states we assimilate them to ourselves.

This view – what I will call the *crystallization thesis* – is compatible with an antirealist view of Autonomy, but it should be clear that it does not presuppose such a view. Even if realism turns out to be the correct view of Autonomy, as I think it will, my claims about the incommensurability and normativity of Autonomy introduced in the previous section remain true. On the realist view of Autonomy, the crystallization thesis serves as an explanation of the prevalence of the cognitive metaphor given the epistemic limits we face in ascribing Autonomy.

The crystallization thesis is intended to apply quite generally: to pedagogical uses of metaphor, to metaphors that create new ways for scientists to think about the organization of nature, to past and present metaphor, to metaphor in science and beyond its bounds. The prevalence of the cognitive metaphor in the biological sciences likely reflects the ease with which other living agents can be assimilated to psychological agents such as ourselves, something even more pronounced now, 150 years after Darwin, than it was prior to the theory of evolution. But it is important to see that the crystallization thesis, if true, also holds in other cases of agency. With that in mind, I turn briefly to consider nonbiological agents.

3. THE CRYSTALLIZATION THESIS AND NONBIOLOGICAL AGENTS

There are three types of agents beyond biology that I shall discuss: artifacts, nonliving natural agents, and nonphysical agents. In each case, the reliance on a cognitive metaphor makes these agents not only more organismlike, but more like a particular kind of organism: human agents. This supports the crystallization thesis insofar as this is just what one would expect were that thesis true, given that the cognitive metaphor can be readily ascribed to the full range of agents, and not just biological agents.

The most pervasive, contemporary use of the cognitive metaphor in describing the workings of artifacts and explaining their behaviors is to computers and the software programs that run on them. They are frequently said to make decisions, to know some things and not others, to understand or not understand what we want them to do, to remember previously presented information, to try to achieve certain goals (from a computer's rebooting to a chess program's attacking a weak Queen-side defense), and in light of their friendly question prompts (such as "Do you really want to delete that file?"), to want users to do or at least be aware of certain things.

Earlier I talked of adopting the agentive stance toward certain chunks of the biological world, with a nod to Dennett's notion of an intentional stance. Dennett himself has emphasized that the intentional stance is an ineliminable strategy for predicting and explaining behavior, and its application to artifacts such as computers is one that has been prominent in his defense of this idea. My point is not one about prediction and explanation but one about the role of the intentional stance in shaping how we conceptualize the kind of agent these artifacts are.

In characterizing agents as causes in the introductory chapter, I mentioned that natural agents, such as hurricanes or storms, are sometimes personified, and that our notion of an agent extended beyond the physical realm to include agents such as gods, angels, ancestral spirits, and ghosts. In the former case, the tendency to adopt the cognitive metaphor in Western societies has diminished radically in the last few hundred years, and what we might think of as the richness of the cognitive attributions made has never been great. But in the latter case, nonmaterial agents are often ascribed extremely rich psychological profiles. By virtue of this, they come to be treated as very much like human agents but exaggerated in their powers, failures, or limitations. They are viewed as able to react to what we do, and how we perceive or treat them, much in the way that human agents can, and to be aware of or even preoccupied with our own lives. These agents take on many of the properties that human agents have, and the psychological properties ascribed to them play a central role here.[4]

The effect of the cognitive metaphor in all three cases, but particularly in those where it runs more deeply – those of artifacts and nonphysical agents – is to heighten our sense of the agency involved. It provides a way of conceptualizing how these individuals function as agents by assimilating them to our paradigm agents, human beings. In short, it crystallizes agency, much as I am claiming that it does in the domain of life.

An interesting question (whose detailed discussion I leave for another occasion) concerns the psychological basis for the use of the cognitive metaphor, and why reliance on it crystallizes agency. I have said that the cognitive metaphor functions by assimilating other living things to us – paradigm cognitive, organismal, and living agents. And we might seek to develop a robust psychological account that takes us beyond this suggestion by appealing to ways in which similarity metrics, metaphors, and paradigms interact to govern cognition. But I suspect that such an account would only scratch the surface to this question. For remaining would be the further question of just why it is that a cognitive metaphor should have this effect or, to put it the other way around, why we should so readily see cognition when there is none there to be seen.

This last way of expressing the issue invites exploration of a different aspect of cognitive psychology: that focused on the "naturalness" of the perception of agency, animacy, and causation, and the relations between them. For we also often see causation when there is none to be seen. Two sets of classic studies are suggestive here: those of causation and action by the Belgian psychologist Albert Michotte, and those of Heider and Simmel's studies of the "apparent behavior" of simple animations. Michotte showed that perceptions of causation were heightened when motion was animatelike, implying that this kind of apparently self-generated motion served as an attractor for our ability to detect causation. Heider and Simmel were able to elicit a variety of judgments of psychological, social, and interpersonal relations between simple geometrical shapes in motion. In both cases motion, or motion of a certain kind, provides the chain linking causation and agency. The real question is where, and how, psychological properties fit into this chain.[5]

One final suggestion here. As has been recognized at least since Darwin's appeal to natural selection as a mechanism based on an explicit analogy with that of artificial selection in the first chapter of *On the Origin of Species*, much of our thinking about organisms turns on an analogy with artifacts. Organisms are treated as if they were designed by an agent to perform certain functions, and their phenotypic features reflect their suitability for achieving those functions. This sort of teleological agency, like causal and intentional agency, comes naturally both to children and to untutored adults. While it is reflected in an evolutionary biology shaped by the theory of natural selection, it clearly predates the Darwinian revolution in natural history. Developmental psychologists disagree about the precise relationship between psychological and teleological agency in the child's developing view of the world, but it seems likely that the two are closely intertwined and mutually reinforcing.[6]

4. CAPTURING ORGANISMAL DIVERSITY

One criterion for an adequate view of organisms, implicit in this chapter and the last, is that it elucidate the relationships between organisms, agency, and life. The Tripartite View does well by this criterion, but it is important to show that the Tripartite View also captures much about both naive and informed views of what things in the world are organisms. As the examples at the outset of the previous chapter indicate, our intuitions here are not always clear. Naive and biologically informed views sometimes part ways, and biologists themselves can disagree about particular cases. Perhaps more important than the answer given to particular instances of the question "Is X an organism?" are the reasons for this answer, particularly as we move from clear to less clear cases. In this section I discuss five relatively easy cases, some of which we have met in passing. I shall aim to show that the Tripartite View is suited to capturing the diversity in these agents of the living world, and to meaningfully demarcating them from both the merely living and the merely biological.

First, the Tripartite View applies readily to organisms in both the plant and animal kingdoms. These are the two kingdoms of the biological world that supply us with the multicellular life forms with which we are most familiar, and it should be clear that the individuals they subsume are living agents that satisfy both Life Cycles and Autonomy. The naive, untutored tendency to view animals as better examples of organisms than are plants, or to see them as, in some sense, more complete agents, is explained by the Tripartite View by reference to the cognitive metaphor and its role in heightening Autonomy, and so crystallizing agency. (Here is one place where motion clearly plays a role.)

Second, the same general point holds true of all six kingdoms of organisms: the Archae, Eubacteria (both prokaryotes); and the Protista, Fungi, Plantae, and Animalia (all eukaryotes). Individuals in all of these satisfy all three parts of the Tripartite View, and while none are any more or less organisms than the others, our views here are sometimes swayed by the cognitive metaphor. Importantly, single-celled and multicellular organisms fall on one side of the divide between organisms and merely living things, with cells that have been integrated into multicellular organisms falling on the other side. I shall return to this point in taking up less easy cases in the next section.

Third, sexually reproducing individuals, despite the fact that they do not make copies of themselves, are organisms, while fragments of DNA, or even whole genes, are not, despite the fact that they are reliably copied or replicated. I have focused on Life Cycles in discussing the difference between these two cases, but in fact, the differences between the two extend to Living Agents and Autonomy as well. Even though genes are biological agents, they are not living kinds, and, like other parts of organisms, have surrendered much of their minimal functional autonomy. As I shall argue in Part Three in discussing genetics and organismal development, again the cognitive metaphor has played a significant role in crystallizing the agency of genes, making them appear more like organisms than they in fact are.

Fourth, individuals that do not themselves reproduce, or even that have lost the physiological capacity to do so, are organisms on the Tripartite View. Just as Life Cycles shifts the onus from mere replication to reproduction, it also picks out having an intergenerational life cycle as a property of some members of a lineage, rather than as a property of every individual in that lineage, as what is relevant to being an organism. So sterility, even obligate sterility, is no barrier to being an organism, and need not be seen as a special or deviant case.

Finally for now, consider kinds of individuals that have changed their status as organisms, such as mitochondria. On the Tripartite View, these remain living agents, but no longer satisfy either Life Cycles or Autonomy. Their existence suggests other cases, such as some cases of obligate parasitism or endosymbiosis, which can be viewed as being transitions from organism to nonorganism. Again, this is important in thinking about examples that are harder cases for any account of organism, agency, and life.

5. CORPORATE ORGANISMS

The Tripartite View provides the resources to capture and illuminate much of the diversity amongst organisms, and so meets one criterion for an adequate view of organisms. But how does it cope with some of the harder cases out there? Recall the three cases with which I began – the humungous fungus, the mother-fetus-placenta triad, and living coral reefs. Part of the difficulty in thinking decisively about each of these examples derives from a feature that these examples share not only with one another but also with a large range of harder cases. These include superorganismal insect colonies, multispecies individuals (such as lichens, composed of fungi, algae, and bacteria), and parasite-host systems. Call these (putative) *corporate organisms.*

The problem that corporate organisms represent can be expressed neatly in terms of the Tripartite View as follows. Although there is reason to think that each of these agents satisfies Life Cycles and Autonomy, as well as (though perhaps more questionably) being living agents, putative corporate organisms, unlike individual organisms, are composed of agents that themselves also satisfy each of these conditions. The introduction of corporate organisms creates a poignant version of what I shall call the *Who's Zoomin' Who Problem*: who is really in control of the various processes that two related or connected organisms participate in? The Who's Zoomin' Who Problem also applies to organismal pairs that are not related mereologically, but I want to focus on just that case in what follows.

Mereological relationships between living agents per se do not generate the Who's Zoomin' Who Problem. Living agents lodged one inside the other do not, for the most part, struggle for control, but have averted or resolved it, typically in favor of the larger agent. Indeed, one might think that this is entailed by the mereological, hierarchical organization of living agents. We have seen that the paradigm of living agents that are not organisms are organs and other parts of organisms, and the sort of control that organisms exert over their parts is captured in the Tripartite View by the condition of Autonomy. Matters are different, however, when organisms themselves are viewed, in effect, as parts of larger organisms.

There is a prima facie tension in an agent itself satisfying Life Cycles and Autonomy, and being a part of another living agent that also satisfies these conditions. For, to put it intuitively, how can it have a life of its own if what it does is subject to the control of some larger agent that has its own life? To put this the other way around, how can a corporate organism maintain its minimal functional autonomy if it contains physical parts that are not subject to its control but exert their own control? These rhetorical questions should make it clear how the Who's Zoomin' Who Problem arises with respect to Autonomy. But much the same problem also exists vis-à-vis Life Cycles. Before turning to that, consider a parallel between biological and cognitive autonomy.

I have been suggesting, in effect, that minimal functional autonomy is a zero-sum game when played between mereologically related living agents. Thus, if there are corporate organisms, their existence comes at the expense of the organismal status of the individuals that constitute them. In introducing the cognitive metaphor at the end of Chapter 2, I mentioned the idea of a group mind that has been recently revived in both the biological and social sciences. In both of these contexts there is recognition of much the same sort of trade-off between the existence of group and individual minds. For founders of the group-mind hypothesis in the social sciences, such as Gustav Le Bon, a group mind came to replace individual minds, just as "the crowd" usurped the individual as an autonomous unit of action. Likewise, contemporary proponents of the group-mind hypothesis in the biological sciences, such as David Sloan Wilson, focus on cognitive adaptations that involve a shift of cognitive control from individuals to groups.[7]

Consider now Life Cycles. Given that living agents are hierarchically organized, the replication of any living agent also involves the replication of the living agents it contains, as well as at least part of the living agent it constitutes. In articulating Life Cycles we saw that reproduction rather than (mere) replication was appropriate in formulating criteria for individuating organisms in the living world, and I used the distinction between reproduction and replication as the basis for arguing that parts of organisms did not satisfy Life Cycles. Parts of organisms do not have their own life cycles, but constitute part of the life cycle of those organisms.

Parity of reasoning may suggest that, in at least some cases of corporate organisms, individual organisms simply form part of their life cycle. Regular organisms are in effect organs of the corporate organisms they physically constitute. Indeed, those defending the superorganismal conception of insect colonies, for example, sometimes explicitly draw on the analogy between the relationship between organs and organisms, on the one hand, and organisms and insect colonies, on the other. Given that insect colonies and other putative corporate organisms, such as lichens and parasite-host systems, not only reproduce but form lineages, what reason is there to deny that they themselves satisfy Life Cycles? But if they do, then they are the agents that form lineages with Life Cycles, and the organisms that constitute them are merely being replicated derivatively.[8]

Questions about just what living agents satisfy Life Cycles not only arise with respect to organisms but in thinking about the agents of life more generally. Some of these questions concern the notions of inheritance and development, and we thus return to something like Life Cycles in Part Three in discussing genetics and ontogeny in detail there. Others concern what sorts of biological agents form lineages and function as agents of natural selection, our focus in Part Four. As we will see, some paleontologists, such as Stephen Jay Gould and Steven Stanley, view *species* as the best candidates for corporate organisms that, together with regular organisms themselves, are the primary agents of selection.[9]

I have indicated how proponents of corporate organisms have sought to defend their views by pointing to the ways in which such organisms share the features that individual organisms have that are specified by Life Cycles and Autonomy. Yet if corporate organisms do strictly satisfy Life Cycles, then they must themselves be living things, and I think that here proponents have been far too complacent in simply assuming corporate organisms to be living agents. Earlier work on superorganisms within the Chicago School of ecology, chiefly by Warder Clyde Allee, Thomas Park, and Alfred Emerson, had concentrated on the *physiology* of populations. But much of this has been put to one side in contemporary defenses of corporate organisms. These have focused on groups as agents of selection, and in self-consciously distancing themselves from earlier traditions in which an appeal to group selection played a central role, they have worked with a notion of a group significantly broader than that of a corporate organism.[10]

The physiologist Scott Turner's advocacy of "the extended organism" represents perhaps the most direct revival of this appeal to physiology and to related notions (such as metabolism, energy transfer, and homeostasis). Turner's views are interestingly ambivalent vis-à-vis individual and corporate organisms, and I want to conclude this section by linking this ambivalence to two ways in which we can think of the individual and biology.[11]

On the one hand, the central claim that Turner seeks to defend is that individual organisms can, and typically do, have a physiology that extends beyond the physical boundaries of their bodies. Turner views this as an adaptation of Richard Dawkins's idea of the extended phenotype. What Turner does, in effect, is extend the reach of the individual organism, just as the extended phenotype, in Dawkins's hands, extends the reach of the selfish gene. To draw on our toolkit of concepts from Chapter 2 again, an individual organism's physiology has a wide realization. This is a form of externalism about physiology, what I have elsewhere called *locational externalism*, because it views the relevant system – in this case, the physiological system – as physically bounded by locations outside of the organism. On this view, the individual organism remains very much the focus of biological investigation, but we move beyond its physical boundaries in order to understand how it functions. Since the organism's physiology is locationally external, its physiological properties do not supervene on the intrinsic, physical properties of the organism. They are not individualistic.[12]

On the other hand, and as suggested by Turner's example of the living coral reef that we discussed at the outset of Chapter 3, we might also consider that a living agent, which incorporates individual organisms, has its own physiology. It can be thought of as a corporate organism in its own right. But then *its* physiology is not locationally externalist, for that physiology is located entirely within the boundary of the corporate organism that has it. On this view, biologists need to frameshift from individuals – indeed, from individual organisms – to corporate organisms, but they can remain individualistic about the properties of these newfound individuals. Just as adopting an extended physiology accentuates the role of the individual organism within biology by rejecting individualism, endorsing corporate organisms reduces that role by shifting from individual to corporate organisms as the relevant agents.

I began this section with the question of how well the Tripartite View coped with harder cases, and I have focused on a range of these concerning putatively corporate organisms. My chief aim here, however, is not to use the Tripartite View to resolve the Who's Zoomin' Who Problem, either in general or in particular cases, or to determine whether (and if so, just when) corporate organisms really exist. Rather, I have tried to show that the Tripartite View has the resources to articulate the problem that corporate organisms represent, and allows us to see more clearly what is involved in positing and defending the existence of corporate organisms.

The final sort of exploration of the Tripartite View that I shall undertake here concerns its implications for some influential views within more speculative areas of biology: that of the early history of life and artificial life. In both areas, claims have been made about organisms that seem at best incautious in light of the Tripartite View.

6. THE PRIORITY OF ORGANISMS AND THE EARLY HISTORY OF LIFE

Amongst today's living agents, organisms loom large – large enough that it is questionable whether there would be a recognizable living world without them. Yet I want to suggest that organisms are not simply prevalent

TABLE 4.1. *The major transitions of life*

Replicating molecules	→ Populations of molecules in compartments
Independent replicators	→ Chromosomes
RNA as gene and enzyme	→ DNA + protein (genetic code)
Prokaryotes	→ Eukaryotes
Asexual clones	→ Sexual populations
Protists	→ Animals, plants, fungi (cell differentiation)
Solitary Individuals	→ Colonies (nonreproductive castes)
Primate societies	→ Human societies (language)

Source: Redrawn from Figure 1.2 of John Maynard Smith and Eörs Szathmary, *The Major Transitions in Evolution* (New York: Oxford University Press, 1995), page 6.

in the biological world today, or in the last few billion years, but always have been. In some sense they were – and continue to be – the sine qua non of life.

Such a view does not sit well with replicator-centric views of the history of life that have gained currency in part through the recent important work within theoretical biology of John Maynard Smith, Eörs Szathmáry, and Richard Michod. Following the work of Leo Buss that recognized that individuality itself was an evolutionary achievement, these authors have sketched a view of the "major transitions in evolution," beginning with molecular replicators and ending with human societies (see Table 4.1). On these views, the history of life is the history of the construction of more complicated life forms from simpler forms of life, and replication is a property of the earliest living agents. These views are significantly different from one another in the details. Buss's focus, for example, was on the transition from cells to multicellular organisms, and he drew on a multilevel view of selection that was explicitly opposed to viewing genic selection as the primary mode of evolution. Maynard Smith and Szathmáry devote most of their attention to other transitions in the evolutionary hierarchy, and disagree with the core details of Buss's account of the transition to multicellularity. Michod views the cell as the "first true individual," appealing to the sort of multilevel theory of selection that Buss relies on but also emphasizing the role of novel levels of genetic fitness. Each of these views is quite sweeping, and in posing a challenge to them here I need to specify the part of them that I mean to take issue with.[13]

The general idea of replicator-first views of the history of life is that replicators, entities that can reliably replicate their structure over time, precede and are ancestors to entities that have a broader range of

structures and functions. Replicator-first views are typically contrasted
with "metabolism-first" views, which take the formation of a stable, internal economy within some bounded biological agent to be the earliest achievement of living things. One reason for replicator-first views is
that replication is a feature of the chemistry underlying amino acids and
proteins, the basic building blocks of living systems. So, the idea that
there were nonliving replicators before there was life is plausible. But as
Maynard Smith and Szathmáry concede in passing, there are also abiotic forms of metabolism, and it is plausible to view these as providing
the monomers from which more complicated polymers (and eventually
amino acids) were built. The fact that there is replication before life
is no more relevant to the question of the temporal priority of replicators over organisms than is the fact that there is a form of metabolism
before life. The question I am interested in here is not whether replication or metabolism "came first." Rather, it is whether there was a period of nonorganismal replication *in the living world.* When living agents
appear, are they initially replicators and later organisms? This question
takes us from the origins of life domain to that of the early history of
life.[14]

It is here that the Tripartite View helps to highlight some problems
for the replicator-first view of that early history. Since it is reasonable to
assume that the first living agents were significantly simpler than present
living things, we should not expect them to instantiate as elaborate a
form of the HPC kind living agent as today's living agents. Nor should we
require too much by way of a life cycle and minimal functional autonomy.
So, if the Tripartite View is to be useful in this context, we need a scaled-
down version of it, one that recognizes a less rich homeostatic property
cluster as determining "living agent," and with less demanding versions
of both Life Cycles and Autonomy.

In his thought-provoking discussion of the evolution of multicellular
life in the final chapter of *The Extended Phenotype*, Richard Dawkins has
expressed just the sort of view of the early history of life that I mean to
take issue with. Dawkins says

The integrated multicellular organism is a phenomenon that has emerged as a
result of natural selection on primitively independent selfish replicators. It has
paid replicators to behave gregariously. The phenotypic power by which they
ensure their survival is in principle extended and unbounded. In practice the
organism has arisen as a partially bounded local concentration, a shared knot of
replicator power.

On this view, the existence of replicators, small genetic fragments, is historically fundamental, and that of (multicellular) organisms derivative. For Dawkins, we need to understand "why replicating molecules ganged up in cells," and why "cells gang[ed] together into multicellular clones" in order to explain why it is that organisms exist at all.[15]

Stromatolitic fossil evidence gathered over the last twenty-five years, particularly by J. William Schopf and his colleagues, indicates that life appeared on Earth at least 3.5 billion years ago, and perhaps as long ago as 3.8 billion years ago (see Figure 4.1). This is significantly earlier than previous estimates of when life began on Earth, and importantly for our purposes, the evidence indicates the presence of membrane-lined structures that seem likely to instantiate a nascent form of the cluster of properties specified by the HPC kind living agent. Stromatolitic fossil mats differ little over the 1.5 billion years of life that precede the evolution of eukaryotic cells (see Figure 4.2), and the process of cyanobacterial photosynthesis was likely present earlier than 3.5 billion years ago. Schopf himself makes a number of claims about the first billion or two years of evolution based on his reading of the stromatolitic evidence that I have discussed elsewhere, and some have raised questions about whether stomatolites are evidence of living agents at all. My interest here is in whether such fossils are evidence of organismal life, given that they are evidence of life at all.[16]

We can state this issue and outline my position on it in terms of the Tripartite View. If the oldest stromatolitic fossils are the remains of living agents that are not organisms, such living agents must have failed to satisfy either Life Cycles or Autonomy. But in this context, either of these failures would also be incompatible with their status as replicators. Thus, the idea of a replicator-first, organism-second view of living agents involves a contradiction.

To be a living *agent*, an entity must be a physically bounded locus of action. To be a *living* agent, it must instantiate the corresponding HPC kind. On the Tripartite View, the only way to be a living agent but not be an organism is to fail to satisfy either Life Cycles or Autonomy. If the first living things were active replicators, and not just entities that were replicated, then surely they satisfied Autonomy, having internal control and external freedom, with respect to their primary property, that of being able to replicate. So nonorganismic early life must fail to satisfy Life Cycles. But how could living replicators that (eventually) produced us fail to be members of a lineage with an intergenerational life cycle? In keeping

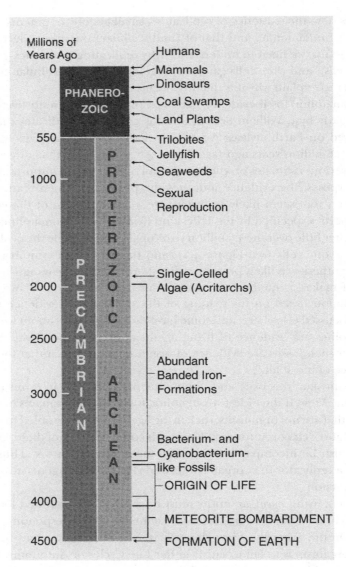

FIGURE 4.1. Life's timeline (Redrawn from Figure 9.8 of J. William Schopf, *Cradle of Life* [Princeton, NJ: Princeton University Press, 1999], page 253.)

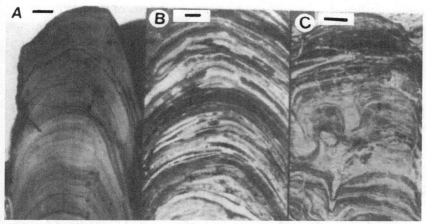

FIGURE 4.2. Stromalitic mats from 1.3, 2, 3.35 BYA (Modified from Figure 7.8 of J. William Schopf, *Cradle of Life* (Princeton, NJ: Princeton University Press, 1999), page 199.)

with the scaled-down version of the Tripartite View appropriate in thinking about simpler living agents, original replicators need not be thought of both as progenerating and developing (that is, reproducing, in Griesemer's terms). But however replication is fleshed out, it is paramount to the replicator-first view that they do form lineages, and that they do so through their own agency. Thus, it is hard to see how early life could fail to satisfy some version of Life Cycles. But then they are not only living agents (by hypothesis), but must also be organisms.

The point this argument makes is simple. Living agents are physically bounded entities with a cluster of structural and functional features. The only nonorganismic living things that we know of in today's world are dependent on the existence of organismic life, and so could not have predated them in the history of life. Minimal functional autonomy and belonging to a reproductive lineage whose members instantiate a life cycle are the key properties that distinguish organisms from the merely living. It seems implausible that replicators that were antecedent to organisms as we know them lacked either. Thus, it is difficult to make sense of preorganismal living agents. To suggest that (living) replicators constructed organisms is to overlook that the first living replicators must themselves have been organisms. Replicative life had its price: organismic existence.

In the broader scheme of the debate between replicator-first and metabolism-first views of the history of life, this conclusion implies that replicator-first views are plausible, at most, as views of the prehistory or origins of life. Perhaps that is all that proponents of replicator-first views have intended. But if so, then their views tell us very little, if anything, about that long stretch of life over the first billion or two years before we see multicellular organisms.

As a coda, it seems to me that the most promising challenge to the general conclusion of the importance of organisms in the early history of life lies in radically reimagining the forms that life cycles may have taken in the early history of life. A development of a speculative line of thinking that the chemist-turned-microbiologist Carl Woese has offered is an interesting alternative compatible with the aspect of replicator-first views that I have been challenging in this section, and I close this section with a brief discussion of it.

Woese's view is located against the background of his earlier advocacy of the three domains of *Archaea, Bacteria,* and *Eucarya* as representing truly phylogenetic taxa that reflect the history of life, a view increasingly accepted since it was first proposed. Woese argues that the universal ancestor to these three genealogical groups was not an organism. This is because in the early history of life there were no organismal lineages, due in part to the prevalence of something like horizontal gene transfer. Woese's hypothesis is that the universal ancestor of all three domains of life, and so of life itself, is not an individual but a community of cell-like structures, progenotes, that evolves as a biological unit. As Woese says, it was "a community of progenotes, not any specific organism, any single lineage, that was our universal ancestor – a genetically rich, distributed, communal ancestor." On this view, natural selection gradually winnows out the variety of progenotes and limits horizontal gene transfer to form living agents with a genealogical lineage that are more like organisms.[17]

As Woese recognizes, replicative mechanisms, such as translation and transcription, evolved as a part of this process, allowing proteins to be processed and so elaborating the metabolism of progenote communities. But Woese begins by assuming the presence of replication, and views his universal ancestor as a living thing of some kind, just not an organism. The idea that early, membraned living things drew on a shared pool of both metabolic and replicative resources, or even that they shared in life cycles, provides one way to rethink the historical relationships between organisms and replication that challenges the conclusions for which I have been arguing in this section.

So in advocating a view in which organisms are seen as the product of processes that involve replication in living agents, rather than being those very agents, Woese's view counts as a replicator-first view of evolution of the type that I have expressed skepticism about. But since Woese seems to envisage the fundamental units on which natural selection operates during the early history of life to be communities of progenotes, rather than individual replicators – the universal ancestor is, after all, a *communal* ancestor – his view involves a fairly radical departure from the sort of smallism to be found in standard replicator-first views. If Woese's thinking here takes us down the right track, then the first organisms evolved from larger biological agents as gene pools started to close and vertical gene transfer became a viable competitor to horizontal gene transfer. Living agents precede organisms, but such agents are not much like the small parts of organisms invoked as part of the gene's-eye view of evolution.

7. ARTIFICIAL AND NATURAL LIFE

Early life forms may have "ganged up" to form more complicated forms of organisms, but I have argued that these initial forms were already organisms, not mere replicators from which organisms derived. Replication is clearly an important feature of the biological world. But just as I have argued that proponents of the gene's-eye view are mistaken in the role they have ascribed to replication in characterizing living agents, so too have they overstated the place of replication in the early history of life. The replicator-centered view of life has also been predominant in work in artificial life, and I want to end the chapter by considering some of that work in light of the argument thus far.

Artificial life explores the prospects for simulating or creating life, typically by using computational models. What researchers on artificial life have begun to do, in effect, is to tease apart the cluster of features that appear together in organisms, experimenting with simulated entities that possess just a few of those corresponding features. Or, to put it in terms that reflect how that research proceeds, A-Lifers are in the business of creating entities that have a subset of the properties that feature in the HPC natural kind living agent, manipulating each of these properties experimentally to explore what happens over time.

Like the core of work in artificial intelligence, much of that in artificial life involves simulating the features and processes deemed central to a given natural phenomenon: intelligence in the one case, and life in the other. In terms of the trichotomous taxonomy of features of living agents introduced in the previous chapter – structural, functional, and wide functional – there has been a concentration in both AI and in A-Life on the second of these, the functional properties. An appeal to the doctrine of "substrate-neutrality" both in AI and in A-Life has given reason to ignore structural properties. This allows the fact that such entities are realized in computer electronics to be seen as irrelevant to what their investigation may tell us about evolution by natural selection in the living world. The neglect of wide functional properties has been legitimated by the naturalness of modeling traits of organisms themselves, and seeing how they vary with parameter changes in the models. This focus and the corresponding neglect lies behind some of the more striking claims made within the A-Life literature.

A number of these concern *Strong A-Life*, the view that it is possible to create living things through the programs that A-Lifers write. One of the founders of A-Life, Chris Langton, introduces Strong A-Life as a central part of the rationale for work in A-Life more generally:

The ultimate goal of the study of artificial life would be to create 'life' in some other medium, ideally a *virtual* medium where the essence of life has been abstracted from the details of its implementation in any particular model. We would like to build models that are so life-like that they cease to become *models* of life and become *examples* of life.[18]

Likewise, Thomas Ray says that the "intent of this work is to synthesize rather than simulate life," and J. Doyne Farmer and Alletta d'A. Belin state that within "fifty to a hundred years, a new class of organisms is likely to emerge," saying that although they will be artificial "they will be 'alive' under any reasonable definition of the word."[19]

Defenders of this sort of view face many of the same challenges as do proponents of strong AI. These turn in large part on their focus on simulating just the functional properties in their respective domains. The issue that I want to raise is whether such a focus itself could ever produce entities that had, and not simply simulated the having of, the properties associated with intelligent or living things.

The functional property of living agents that has received the most attention within artificial life research is that of the capacity for reproduction or self-replication. This is so not only within the frameworks of genetic algorithms or evolutionary computation, but more generally. For example, Dawkins's biomorphs, Daniel Hillis's coevolutionary simulations, and Thomas Ray's Tierra – three widely discussed examples of early A-Life work – all share this focus. A-Lifers speak freely of how their creatures or "digital organisms" fare in the process of evolution by natural selection. I propose that we take Strong A-Life seriously and use the Tripartite View to see what this talk of digital organisms amounts to.[20]

On the Tripartite View, an organism instantiates the HPC kind living agent and satisfies both Life Cycles and Autonomy. If we insist on a strict understanding of Life Cycles, so that anything satisfying it both progenerates and develops, this would limit the ways in which replication could be carried out within software programs. But it need not rule out the possibility of nonbiological agents literally forming intergenerational lineages in which reproduction occurs. Thus, there seems no problem in principle in artificial life satisfying, rather than merely simulating the satisfaction of, Life Cycles.

We might adopt the same view of Autonomy. I have characterized minimal functional autonomy in terms of the relationship between control and the boundaries of the entity: internal control within its boundary, and freedom from control beyond its boundary. The naive position on the autonomy of artificial life critters is that, unlike biological organisms, they have at best *derived* autonomy, in much the way that the intentionality that AI programs have is typically thought to have some sort of derivative status. Biological organisms have *real* or *original* autonomy, while artificial life critters have derived autonomy. On the Tripartite View, Strong A-Life is committed to challenging this view. Current digital organisms may not have a very high level of minimal functional autonomy, but one of the goals of Strong A-Life is to develop programs that increase the number and range of dimensions in which they exercise minimal functional autonomy. This property, one might think, is just like other functional properties, and being independent of how it is realized, could be realized in nonbiological matter.[21]

In light of the incommensurability and normativity of minimal functional autonomy, and the subsequent questions they raise concerning realism, it might seem unfruitful to argue over whether digital organisms "really" satisfy Autonomy or merely simulate the property that it ascribes to them. In any case, the core problems with Strong A-Life lie elsewhere, with whether digital organisms are living agents.

Consider the HPC natural kind living agent. There seem no barriers to thinking that A-Life critters literally possess, as opposed to merely simulate the possession of, many if not all of the functional properties that feature in this homeostatic property cluster, such as growth and repair. More problematic is the possession of both the structural and the wide functional properties in that cluster. Because of their simulated nature, they do not possess the same physical properties as the entities they simulate – indeed, that is seen to be one of the virtues of research in artificial life, much as it is in artificial intelligence. Included here would be properties such as having heterogeneous parts or specialized physical structures. Lacking a body, for example, they do not have body parts. Lacking cells, they do not have organelles. And being realized in silicon chips, they do not possess proteins or enzymes. Likewise, because they typically occupy only a simulated environment, they also can only simulate the wide functional properties that living things have, properties that require interaction with an actual environment, such as niche construction or specific adaptations.

This is not to deny that many, if not all of these properties can be simulated, and that such simulations can be deeply informative about the structures, processes, and events they simulate. But what these properties share is precisely the tight connection to physical embodiment that work in artificial life prides itself on abstracting from, and that is why the program of Strong A-Life represents a confusion.

The parallel with artificial intelligence is again informative. If what is crucial to intelligence is not simply the formation and manipulation of symbols in accord with rules (as in classic AI), or the adjustment of connection strengths through learning algorithms (as in connectionism), but the exploitative representation of, and action in, one's environment, then the very program of Strong AI is a nonstarter. Likewise, if to be not just an organism but a living agent is to have certain physical structures and to interact with one's environment in certain ways, then programs that are physically realized in computers and that interact merely with a simulated environment are not candidates for living agency and the program of Strong A-Life is a nonstarter.

The basic claim here is again fairly simple. There are certain properties for which "to be is to be simulated": replication is like this, as are purely functional properties. There are others for which there is an unbridgeable gap between simulation and possession. Many of the properties that constitute the HPC natural kind living agent are of this latter kind, and given that, digital organisms are not simply not organisms, but not even living agents.

8. THE RESTORATION OF THE ORGANISM

In this chapter, I have taken up some of the intricacies involved in articulating and defending the Tripartite View of organisms. The Tripartite View gives organisms a certain kind of distinctive place amongst living agents, particularly with respect to their physical constituents. As such, it corrects the smallist views of the living world that, in effect, remove organisms from the central place that they have traditionally occupied in thinking about living agents, replacing them with genes as the principal agents of life. I have also addressed what we might see as the "threat from above" – what the existence of corporate organisms poses to the distinctiveness of individual organisms in the biological world. The "threat" here, however, is not one of global displacement but of localized losses of organismal status amongst what one would otherwise think of as individual organisms.

I have also argued that there is an important sense in which organisms are more fundamental units in the history of life than the means by which they propagate their life cycles. In addition, failing to take the bodily nature of organisms seriously has led some researchers in artificial life into confusion about just what their work shows. These conclusions are themselves of limited scope in that they do not (and do not aim to) strike at the heart, respectively, of replicator-first views of the history of life and work in artificial life. But I trust that they both illustrate how the Tripartite View can direct us in probing some of the more speculative claims made in these areas, and to serve as further food for thought.

In the next chapter I shift from our focus on organisms as living agents to another natural kind that has occupied centerstage in thinking about biological hierarchies, that of species. I have already noted that the HPC view of natural kinds has had its primary application to species, but also that the predominant view of species is that they are not natural kinds at all, but individuals. How to resolve the dissonance between this pair of claims will be part of our task in the remainder of Part Two.

5

Specious Individuals

1. SPECIES AND THE LINNAEAN HIERARCHY

When the Swedish naturalist Karl Linné, better known then as now under his Latinized name, Carolus Linnaeus, proposed a system for biological taxonomy in the first half of the eighteenth century, he could hardly have predicted that it would continue to form the backbone for biological classification and taxonomy over 250 years later. Two features of the system that Linnaeus proposed structure systematics today: the idea that biological nature was hierarchically organized, with higher levels in the hierarchy subsuming lower levels, and the Latinized binominal nomenclature that is still used in the naming of biological taxa. Reflecting Linnaeus's own debt to the Aristotelian distinction between species and genus, these binominals name species but also reflect the genus to which the species belongs. Thus, *Homo sapiens* uniquely names our own species and the name tells us that we belong to the genus *Homo*. Species and genus are distinct ranks in the Linnaean hierarchy, and both were taken by Linnaeus to be natural categories, part of the organization of nature itself.

The contemporary Linnaean hierarchy includes many more ranks than just species and genus. Linnaeus recognized higher ranks, such as the *empire* of all things, three *kingdoms* of animal, vegetable, and mineral, six *classes* of animal (mammals, birds, reptiles, fish, insects, and worms), and eight *orders* of mammals. As the examples of taxa within his categories of kingdom and class indicate, some of the particular taxa that Linnaeus proposed are still seen today as natural categories, while others have come to be viewed as the result both of the massively incomplete knowledge that naturalists possessed in the early eighteenth century and

Kingdom

Phylum

Subphylum

Superclass

Class

Subclass

Infraclass

Cohort

Superorder

Order

Suborder

Infraorder

Superfamily

Family

Subfamily

Tribe

Subtribe

Genus

Subgenus

Species

Subspecies

FIGURE 5.1. The expanded Linnaean hierarchy (Modified from Figure 6.1 of Marc Ereshefsky, *The Poverty of the Linnaean Hierarchy* [New York: Cambridge University Press, 2001], page 213.)

of the mistaken theoretical assumptions that lay behind the Linnaean hierarchy. Even so, it is striking that the Linnaean hierarchy itself has primarily been augmented and revised, rather than abandoned, particularly given the radical changes in the nature of views of the biological world in the last 300 years, and our knowledge of it. Figure 5.1 gives a compact representation of the twenty-one ranks usually considered in the full, expanded Linnaean hierarchy, with "+" used to denote an additional superordinate rank and "−" an additional subordinate rank, sharing the same stem as a given rank.

Species and genera are often thought of as occupying a special place in the Linnaean hierarchy, one that imbues them with a significance in evolution that both entities below them – varieties and individuals – and taxa above them – families, orders, classes, and phyla – lack. Lower order entities in this taxonomic hierarchy are not sufficiently distinct, discrete, or permanent to be agents that influence the direction that evolution takes, while higher taxa do not fall under natural categories at all, but under artifactual categories of convenience. Indeed, since varieties or races are individuated by various criteria according to our needs and fancies, there is reason to regard both sub- and superspecific Linnaean categories as lacking the reality that species have. Since the categories species and genus were developed by Linnaeus and other eighteenth-century naturalists from our existing folk taxonomies of the biological world, the claim that "[h]umans everywhere classify animals and plants into specieslike groupings that are as obvious to a modern scientist as to a Maya Indian" is not as striking as it might otherwise appear.[1]

Linnaeus has been a stalking horse for those critical of both essentialism about biological natural kinds and of the idea that the biological world is monistically ordered. In Chapter 1, I identified the idea that species were individuals, not natural kinds, and pluralism about species concepts as alternatives, respectively, to essentialist and unificationist views of species. I shall take up both of these views in the following sections, beginning with how each is motivated by shortcomings of the sort of traditional realist view that underpins the Linnaean hierarchy.

2. REALISM, PLURALISM, AND INDIVIDUALITY

Recall that in Chapter 3, I characterized traditional realism in terms of three assumptions:

> *commonality assumption*: there is a common, single set of shared properties that form the basis for membership in any natural kind
> *priority assumption*: the different natural kinds there are reflect the complexities one finds in nature, rather than our epistemic proclivities
> *ordering assumption*: natural kinds are ordered so as to constitute a unity

Pluralists about species reject at least one of the priority and ordering assumptions. For example, the bryologist Brent Mishler and the botanist Michael Donoghue reject the ordering assumption but maintain the priority assumption when they say that "a variety of species concepts are necessary to adequately capture the complexity of variation patterns in nature." John Dupré, by contrast, would seem to reject both the priority and ordering assumptions in suggesting that "the best way of [classifying species] will depend on both the purpose of the classification and the peculiarities of the organisms in question." Philip Kitcher seems to share this view when he says that "there is no unique relation which is privileged in that the species taxa it generates will answer to the needs of all biologists and will be applicable to all groups of organisms."[2]

Traditional realism about species is indefensible, and in the next two sections I shall indicate just how this has motivated the individuality thesis (section 3) and pluralism (section 4). But reflection on the similarities between the case of species and that of taxonomy in neuroscience leaves me skeptical about the plausibility of the inferences to these two views about species (section 5). Moreover, the resources afforded by the HPC view of natural kinds, introduced in Chapter 3 in characterizing living agents, provide a view of species that lies between traditional realism, on the one hand, and the individuality thesis and pluralism, on the other (section 6). I suggest that rather than rejecting the connection within traditional realism between realism, essence, and kind, we need to complicate those relationships in a way that leaves one closer to traditional realism about species than one might have expected.

3. INDIVIDUALITY AND SPECIES TAXA

A natural way to apply traditional realism to species would be to hold that members of particular species share a set of morphological properties, or a set of genetic properties, each necessary and together sufficient for membership in that species. Let me take the morphological and genetic versions of this view separately. For example, on the former of these views, domestic dogs, members of *Canis familiaris*, share some set of observable properties, such as having four legs, hair, a tail, two eyes, upper and lower teeth, each necessary and together sufficient for their being members of that kind. These are the essential properties of being a member of *Canis familiaris*. On the latter of these views, it is not these morphological properties themselves that form the species essence, but the genetic properties, such as having particular sequences of DNA in the genome, causally responsible for those morphological properties that do so. In either case, the idea is that there is some set of intrinsic properties, the essence, that all and only members of *Canis familiaris* share, whether this be the sort of morphological property that can be readily observed (and thus available to both common sense and science) or the sort of genetic property whose detection requires special scientific knowledge of a more theoretical sort. The question answered by those who posit phenotypes or

genotypes as essences is this: what are the phenotypic or genotypic properties that an individual must have to be a member of a given species S?
The answer to this question, in turn, allows one to answer the question
of what distinguishes S from other species.

The chief problem with either suggestion is empirical. In investigating the biological world, we do not find groups of organisms that are
intraspecifically homogenous and interspecifically heterogeneous with
respect to some finite set of phenotypic or morphological characteristics. Rather, we find populations composed of phenotypically distinctive
individual organisms; sexual dimorphism and developmental polymor-
phism are just two common forms of phenotypic variation within species.
There simply is no set of phenotypes that all and only members of a given
species share. This is true even if we extend the concept of a phenotype
to include organismic behavior amongst the potentially uniquely identifying properties that mark off species from one another. Precisely the
same is true of genetic properties. The intrinsic biological variability or
heterogeneity of species with respect to both morphology and genetic
composition is, after all, a cornerstone of the idea of evolution by natural
selection.

The emphasis on morphology and genotypic fragments as providing
the foundations for a taxonomy of species is also shared by pheneticists
within evolutionary biology, though their strident empiricism about taxonomy would make it anachronistic to see them as defending any version of realism or essentialism. (We might see pheneticism as an attempt
to move beyond traditional realism about species by shedding it of its
distinctly realist cast.) The idea of pheneticism is that individuals are
conspecifics with those individuals to which they have a certain level of
overall phenetic similarity, where this is a weighted average of the individual
phenotypes and genetic fragments individual organisms instantiate.[3]

Both the traditional realist view of species and pheneticism focus on
shared phenotype or genotype as the basis for species membership. The
problem of intraspecific heterogeneity with respect to any putatively essential property that the traditional realist faces is sidestepped by the
pheneticist by, in effect, doing away with essences altogether. However,
the pheneticist still treats species as kinds, rather than individuals. But
they are nominal rather than natural kinds, since the measure of overall
morphological similarity is a function of the conventional weightings we
assign to particular morphological traits or DNA segments.

By contrast, proponents of the individuality thesis respond to the failure of essentialism with respect to species taxa by claiming that species

are not natural kinds at all but individuals or particulars, with individual organisms being (not members of the species kind but) parts of a species. A species is an individual. Species have internal coherence, discrete boundaries, spatio-temporal unity, and historical continuity – all properties that particulars have, but which neither natural nor nominal kinds have. Viewing species as individuals, rather than as kinds, allows us to understand how species can have a beginning (through speciation) and an end (through extinction); how organisms can change their properties individually or collectively and still belong to the same species; and why essentialism goes fundamentally wrong in its conception of the relationship between individual organism and species.

4. PLURALISM AND THE SPECIES CATEGORY

The individuality thesis is a view of the nature of particular species taxa, for example, of *Canis familiaris*. Since I suggested that the individuality thesis was a competitor to both traditional realism and pheneticism, I also think of these views as making claims about particular species taxa. But pheneticism is also often taken as a view about the species *category*, that is, as a view about what defines or demarcates species as a concept that applies to a unit of biological organization. So construed, pheneticism is the view that species are individuated by a measure of overall phenetic similarity, with organisms having a certain level of overall phenetic similarity counting as species, and higher-level and lower-level taxa having, respectively, lower and higher levels of similarity.

Apart from pheneticism, the various proposals that have been made about what characterizes the species category are often divided into two groups. First, there are *reproductive* views, which emphasize reproductive isolation or interbreeding as criteria, including Ernst Mayr's so-called biological species concept, and relaxations of it such as Hugh Paterson's recognition concept and Alan Templeton's cohesion species concept. Second, there are *genealogical* views, which give phylogenetic criteria the central role in individuating species and are typified by Joel Cracraft and Ed Wiley. Unlike pheneticism, both of these families of views fit naturally with the individuality thesis as a view of species taxa. Here is a brief explanation of why.[4]

The focus of both reproductive and genealogical views, as views of the species category, is on two questions: (a) what distinguishes species from other groupings of organisms, including varieties below and genera above, as well as more clearly arbitrary groupings?, and (b) how are particular species distinguished from one another? The question that preoccupies pheneticists, namely, that of what properties of individual organisms determine species membership, receives only a derivative answer from proponents of reproductive and genealogical views. By answering either or both (a) and (b), one determines which species individual organisms belong to not by identifying a species essence, but by seeing which group individuated in accord with the relevant answer to (a) or (b) those organisms belong to. Thus, "belonging to" can be understood in terms of part-whole relations, as it should according to the individuality thesis. Moreover, proponents of reproductive views conceive of species as populations, while proponents of genealogical views conceive of species as lineages, and both of these are easily understood as spatio-temporal, bounded, coherent individuals, rather than kinds, be they natural or nominal.

It is widely accepted that there are strong objections to the claim that any of these types of proposals – pheneticism, reproductive views, or genealogical views of the species category – is completely adequate. And these objections have, in turn, motivated pluralism about the species category, the idea being that each of the three, or each of the more specific forms that they may take, provides a criterion for specieshood that is good for some purposes but not all. The commonality assumption is false because, broadly speaking, phenetic, reproductive, and genealogical criteria focus on different types of properties for species membership, so there is no *one* type of property that determines kind membership. The priority assumption is false, since the different species concepts that one derives reflect the diverse biological interests of (for example) paleontologists, botanists, ornithologists, bacteriologists, and ecologists. Thus, which species concept we use will depend as much on our epistemic interests and proclivities as on how the biological world is structured. And the ordering assumption fails because where we locate the species category amongst other scientific categories depends on which research questions one chooses to pursue about the biological world.

Like pheneticism, reproductive and genealogical views of the species category recognize the phenotypic and genotypic variation inherent in biological populations, and so concede that there is no traditionally conceived essence in terms of which species membership is defined. But even aside from viewing heterogeneity amongst conspecifics as intrinsic to species, these two views share a further feature that makes them incompatible with the sort of essentialism that forms a part of traditional realism. By contrast with the traditional view that essences are sets of intrinsic properties, reproductive and genealogical views of the species category imply that the properties determining species membership for a given organism are not intrinsic properties of that organism at all, but depend on the relations it bears to other organisms.

Although we are considering reproductive and genealogical views of the species category, I said earlier that these views also have a derivative view of what determines species membership for individual organisms. Reproductive views imply that a given individual organism is conspecific with organisms with which it can interbreed, with which it shares a mate-recognition system, or with which it has genetic or demographic exchangeability. Genealogical views imply that conspecificity is determined by a shared pattern of ancestry and descent, or a shared lineage with its own distinctive evolutionary tendencies and historical fate. Rather than conspecificity being determined by shared intrinsic properties, on these views it is determined by organisms standing in certain relations to one another. This is clearest if we consider both views in conjunction with the individuality thesis, since conspecificity is then determined by an organism's being a part of a given reproductive population or evolutionary lineage, where neither of these is an intrinsic property of that organism. Here we depart from individualism about both species category and species taxa, and more generally we seem a long way from the traditional realist's conception of essentialism.[5]

Any serious proposal for a more integrative conception of species must reflect the intrinsic heterogeneity of the biological populations that are species, and it is difficult to see how the traditional realist view of natural kinds can do so. Given the implicit commitment of both reproductive and genealogical views of the species category to an organism's relational rather than its intrinsic properties in determining conspecificity, the prospects for resuscitating essentialism look bleak.

5. BETWEEN TRADITIONAL REALISM, INDIVIDUALITY, AND PLURALISM: THE CASE OF NEURAL TAXONOMY

Our discussion of organisms and living things in Chapter 3 should make us wary of individualistic views of biological categories. In fact, species is not unusual in being a biological category whose members are intrinsically heterogeneous and relationally taxonomized. It seems to me telling that while traditional realism is rendered implausible for these biological categories for much the same reasons that we have seen it to be implausible for species, there is little inclination in these other cases to opt for either an individuality thesis about the corresponding "taxa" or pluralism about the corresponding category. The categories I have in mind and on which I shall focus here are *neural* categories. I want to discuss two of these with an eye to pointing the way to a view of species somewhat closer to traditional realism than might seem defensible, given the discussion thus far.

The first of these is the categorization of neural crest cells. In vertebrate embryology, the neural plate folds as the embryo develops, forming a closed structure called the neural tube. Neural crest cells are formed from the top of the neural tube, and are released at different stages of the formation of the neural tube in different vertebrate species (see Figure 5.2). In neurodevelopment, cells migrate from the neural crest to a variety of locations in the nervous system, the neural crest being the source for the majority of neurons in the peripheral nervous system. Cell types derived from the neural crest include sensory neurons, glial cells, and Schwann cells; neural crest cells also form a part of many tissues and organs, including the eye, the heart, and the thyroid gland.[6]

Neural crest cells are not taxonomized as such by any essence, as conceived by the traditional realist. The category of neural crest cells is intrinsically heterogeneous, and individual cells are individuated, in part, by one of their relational properties, their place of origin. But perhaps the category "neural crest cell" is not itself a natural kind but, rather, a close-to-common-sense precursor to such a kind. (After all, not every useful category in science is a natural kind.) The real question, then, would be by what criteria refined natural kinds that derive from this category are individuated.

I shall focus on the distinction that neuroscientists draw between *adrenergic* and *cholinergic* cells, both of which originate in the neural crest, since this taxonomy of neural crest cells seems initially promising as a candidate for which traditional realism is true. Adrenergic cells produce the neurotransmitter noradrenaline and function primarily in the sympathetic nervous system; cholinergic cells produce acetylcholine, functioning primarily in the parasympathetic nervous system. This truncated characterization of adrenergic and cholinergic cells suggests that they may fit something like the traditional realist view of natural kinds: these two types of neural crest cells are individuated by intrinsic properties or causal powers, their powers to produce distinctive neurotransmitters, which serve as essences determining category membership.

Such an individualistic view of these neural categories, however, would be mistaken, a claim I will substantiate in a moment. But just as mistaken

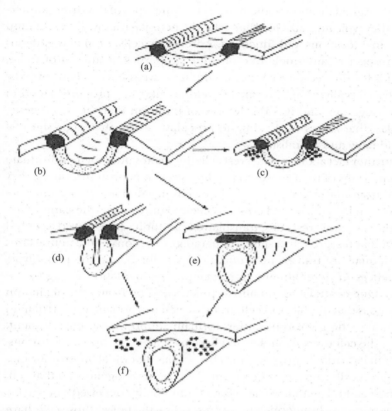

The neural crest. A representation of the localization of the neural crest and neural crest cells (black) between neural ectoderm (stippled) and epidermal ectoderm (white) at neural plate (a), neural fold (b,c) and subsequent stages (d-f) of neural crest cell migration to illustrate patterns of migration in relation to neural tube closure in various vertebrates. The time of initial migration varies between different vertebrates and can also vary along the neural axis in a single embryo. In the rat, cranial neural crest cells migrate while the neural tube is still at the opin neural fold stage (c). In birds, neural crest cells remain in the neural folds until they close (d), only then beginning to migrate (f), whereas in amphibians, neural crest cells accumulate above the closed neural tube (e) before beginning their migration (f).

FIGURE 5.2. The neural crest. (Modified from B.K. Hall and Hörstadius, *The Neural Crest* (Oxford: Oxford University Press, 1988).)

would be the claim that adrenergic cells don't form a natural kind but constitute an individual, or the claim that we should be pluralists about this category of neural cells, claims I shall discuss further in later sections. Standard taxonomic presentations of the two types of cells proceed by introducing a list of features that each cell type possesses, including their typical original location in the neural crest, the types of dendritic connections they typically make to other cells, the neural pathways they take, and their final locations and functions. Adrenergic cells are heterogeneous with respect to any single one of these properties or any set of them, and it is for this reason that they do not have an essence as conceived by traditional realists. Yet in normal development, these properties tend to cluster together, and it is this feature of the form that the heterogeneity takes that allows us, I think, to articulate a view that stops short of individuality and pluralism.[7]

A further fact about neural crest cells dooms any attempt to individuate them in terms of their power to produce certain neurotransmitters: they are *pluripotential* in a sense that I will specify in a moment. Since one goal of research into the neural crest has been to understand the paths of migration of neural crest cells, transplantation studies have played a central role in that research. In a standard paradigm, sections of the neural crest from a quail are transplanted into a chick embryo, and the phenotypic differences in development (for example, pigmentation changes) are noted. One central and initially surprising finding from transplantation studies was that neural crest cells transplanted to a host environment tend to produce the neurotransmitter normally found in that environment, even if the cell transplanted would have produced the other neurotransmitter in its normal environment. This implies that which neurotransmitter a given cell produces is determined by factors exogenous to that cell. The best candidate that we have for a traditionally conceived essence for adrenergic and cholinergic neural crest cells – the power those cells have to produce norepinephrine or acetylcholine, respectively – is not even an intrinsic property of cells. The very property we are supposing to be essential for cell type varies from cell to cell not according to facts about that cell's intrinsic properties, but in accord with facts about the environment in which the cell is located. In terms we introduced earlier, it is a wide disposition, a property of an individual having a wide realization.

As a second example, consider the taxonomy of retinal ganglion cells. These cells receive visual information via the retina and have been extensively studied in the cat and the frog. The neuroscientist Leo Chalupa says that "we now know more about the anatomical and functional properties of retinal ganglion cells than we do about any other neurons of the mammalian brain," suggesting that the neural categories here are the product of relatively well-developed neuroscience. Over the last thirty years, a number of taxonomies have been proposed for retinal ganglion cells, some of these (for example, the alpha/beta/gamma trichotomy) are based on morphological criteria, such as dendritic morphology and axon size, while others (for example, the Y/X/W trichotomy) are based on physiological properties, such as the size of the receptive field (see Table 5.1). The functional distinctness of each of these kinds of retinal ganglion cell suggests that they form distinct visual channels that operate in parallel in visual processing.[8]

As with neural crest cells and their determinate kinds, such as adrenergic and cholinergic cells, the taxonomy of retinal ganglion cells proceeds by identifying clusters of properties that each type of cell has. No one of these properties is deemed either necessary or any set of them sufficient for classification as a Y, X, or W cell, and thus there is no essence for any of these neural categories. But again, I want to suggest that it is implausible to see, for example, groups of Y cells as individuals rather than as a natural kind, or to claim that this way of categorizing retinal ganglion cells has a pluralistic rather than a unificationist basis. The clustering of the various morphological and physiological properties in these cells again points us to a middle ground here. Large numbers of retinal ganglion cells tend to share many of a cluster of properties in their normal environments, and this fact, together with the distinctness of these clusters of properties, provides the basis for individuating retinal ganglion cells into various kinds.

The biological facts in these areas of neuroscience defy philosophical views that posit traditionally conceived essences. But equally clearly they suggest an alternative to the corresponding individuality thesis and pluralism about taxonomy in the philosophy of biology more generally.

6. HOMEOSTATIC PROPERTY CLUSTERS AND THE REVIVAL OF ESSENTIALISM

The middle-ground position I have in mind is based, as I have said, on the HPC view of natural kinds. The philosopher Richard Boyd initially introduced this view as part of his defense of a naturalistic version of realism in ethics, but from the outset he clearly intended it to apply to natural kinds in science and to species in particular. Precursors to the HPC view include Ludwig Wittgenstein's discussion of cluster concepts via the metaphor of family resemblance; Hilary Putnam's introduction of a law-cluster view of scientific concepts; and David Hull's argument that biologists who recognize higher taxa as cluster concepts should extend this view to species themselves. Boyd's previously published discussions have been relatively programmatic, and his most recent view of the

TABLE 5.1. *Some properties of retinal ganglion cells*

	Y Cells	X Cells	W Cells
Receptive field center size	Large, 0.5–2.5°		Large, 0.4–2.5°
Linearity of center-surround summation	Nonlinear	Linear	Not tested
Periphery effect	Present	Usually absent	Absent
Axonal velocity	Fast, 30–40 m/sec	Slow, 15–23 m/sec	Very slow, 2–18 m/sec
Soma size, peripheral retina	Large, >22 μm	Medium, 14–22 μm	Small, <15 μm
Proportion of population	<10%	Approximately 40%	Approximately 50–55%
Retinal distribution	Concentrate near area centralis, more numerous relatively in peripheral retina	Concentrate at area centralis	Concentrate at area centralis and in streak
Central projections	To laminate A, A_1, and C_{12} of LGN, to MIN and, via branching axon, to SC from the A-laminae of LGN to cortical areas 17 and 18, also by branching axon; and from MIN to areas 17, 18, 19	To laminate A, A_1, and C_{12} of LGN; thence to area 17; to midbrain (a minority), but probably not to SC	To SC, to C-laminae of LGN and thence visual cortex area 17 and/or 18, and 19
Nasotemporal division	Nasal cells project contralaterally; most temporal cells ipsilaterally; strip of intermingling centered slightly temporal to area centralis	Nasal cells project contralaterally, temporal cells project ipsilaterally; narrow strip of intermingling cells on area centralis	Nasal cells project contralaterally; most temporal cells also project contralaterally; about 40% of temporal cells project ipsilaterally

Source: Modified from M. Rowe and J. Stone, "Naming of Neurones: Classification and Naming of Cat Retinal Ganglion Cells," *Brain, Behavior, and Evolution* 14 (1977), pages 185–216.

implications of the HPC view for issues concerning species is quite differ-ent from the view I advocate here.[9]

Let me briefly recount the basic features of the HPC view from Chapter 3. At its core is the idea that at least some natural kind terms are defined by a cluster of properties, no particular n-tuple of which must be possessed by any individual to which the term applies, but some such n-tuple of which must be possessed by all such individuals. The mecha-nisms and constraints that cause the systematic coinstantiation or clus-tering of the properties mentioned in HPC definitions act as a form of homeostasis, such that an individual's possession of any one of these prop-erties significantly increases the probability that that individual will also possess other properties that feature in the definition.

The clustering of properties that define HPC kinds, and the fact that this clustering is underwritten by facts about internal organization and external correspondence, marks out the HPC view as a realist view of kinds. On the HPC view, our natural kind concepts are regulated by information about how the world is structured, not simply conventions we have established or language games we play. Before moving to the case of species, consider how the HPC view applies to our pair of neural kinds.

First, take the case of the individuation of neural crest cells. For a cell to be adrenergic is for it to have a certain cluster of properties that sci-entists have discovered; amongst other things, it is to originate in the posterior of the neural tube, follow one of a given number of migra-tory paths, function in the sympathetic nervous system, and produce the neurotransmitter norepinephrine. Facts about the structure of the biological world explain why these properties tend to be coinstantiated by certain kinds of cells. This clustering is the result of incompletely understood mechanisms that govern an embryo's development, and is absent, either partially or wholly, just when the normal function of those mechanisms is disrupted. But no single one of these properties is strictly necessary for a cell to be adrenergic. The presence of all of them, how-ever, is sufficient for a cell to be adrenergic, at least in the environments in which development normally occurs. (Alternatively, one could make the normal developmental sequence part of the homeostatic property cluster definition itself. In either case, the sufficiency is not one that allows for consideration of all logically or even nomologically possible cases.) This feature of the HPC view marks one of the affinities between it and traditional realism, about which I will say more later. On this view, adrenergic neural crest cells are a natural kind of cell of which individual cells are members by virtue of satisfying the HPC definition of that natural kind.

Second, take the case of the individuation of retinal ganglion cells. Consider in particular the physiological taxonomy of Y, X, and W cells. The tendency of the various physiological properties, such as the axonal velocity, soma size, and retinal distribution to be coinstantiated by particular types of cells is no accident, but the result of underlying mechanisms governing neural development and neural functioning. Again, a determinate form of any one of these properties could be absent in a particular cell, yet the cell may still be a Y cell, so no *one* of these properties is an essential property for being a member of that kind of retinal ganglion cell. Still, there is a general definition of what it is to be a Y cell, one based on the homeostatic cluster of properties that one finds instantiated in some cells and not others. Y cells are a natural kind of cell with a sort of essence, albeit one different from that characterized by traditional realism. Moreover, there is a kind of integrity to being a Y cell that invites a unificationist rather than a pluralistic view of it.

There is a more ambitious way to apply the HPC view to this example worth noting. It is a substantive hypothesis that the morphological and physiological taxonomies of retinal ganglion cells are roughly coextensive, but one that is reasonably well confirmed. The HPC view provides a natural way of integrating the two taxonomies by, in effect, adding the two lists of properties in each cluster together. This assumes, of course, that there are common mechanisms that explain the presence of this new cluster of properties qua cluster, without which we would simply have a disjunction of two homeostatic clusters, not a new homeostatic cluster of properties.[10]

I suggest that the HPC view applies to species taxa as follows. Particular species taxa are natural kinds that are defined by a homeostatic cluster of morphological, genetic, ecological, genealogical, and reproductive features. This cluster of features tend to be possessed by any organism that is a member of a given species, though no one of these properties is a traditionally defined essential property of that species, and no proper subset of the cluster is a species essence. (Even properties that are universally shared by all members of the particular species are not essences.) This clustering is caused by only partially understood mechanisms that regulate biological processes, such as inheritance, speciation, and morphological development, together with the complex relations between them. More generally, the homeostatic clustering of these properties in individuals belonging to a single species is explained by facts about the structure of the biological world. For example, organisms in a given species share morphology in part because they share genetic structures, and they share these because of their common genealogy. This is not to suggest, however, that any one of these properties is more basic than all of the others, or that there is a strict ontological hierarchy on which they can all be placed, for the dependency relations between these properties are complex and almost certainly many.

Having severed the connection between the HPC view and traditional realism, let me now indicate an important affinity that the two views share. While possession of individual properties or n-tuples of the relevant homeostatically clustered properties are not necessary for membership in the corresponding species kind, possession of all of them *is* sufficient for membership in that kind. If the homeostatic property cluster definition is sufficiently detailed, this circumstance will likely remain merely an idealization, uninstantiated in fact but approximated to a greater or lesser extent in particular cases. This points to one way in which the sort of essentialism that forms a part of traditional realism is a limiting case of the sort of essentialism that is implicit in the HPC view of natural kinds.

The HPC view can also be applied to the species category, allowing a definition of what sorts of thing a species is that marks it off from other biological categories. First, the general nature of the cluster of properties – morphology, genetics, genealogy, and so on – will distinguish species from nonevolutionary natural kinds such as cells (in physiology), predators (in ecology), and diseases (in epidemiology). Second, species will be distinguished from other evolutionary ranks, such as genera above and varieties below, by the particular specifications of this general cluster of properties. For example, for species for which reproductive criteria are applicable, reproductive isolation will distinguish a species from the mere varieties within it (since the latter are not so isolated), and interbreeding across the population will distinguish it from the genus to which it belongs. In some cases, the distinction of species as a particular rank in the biological hierarchy will be difficult to draw, but I suggest that this is a virtue, not a liability, of the HPC view, since varieties sometimes are very like species (for example, in cases of so-called incipient species), and species sometimes are very like genera (for example, in cases of geographically isolated populations that diverge only minimally and share a recent ancestor).

In defending the Tripartite View of organisms in Chapter 4, I indicated that the HPC view was a realist view. It claims that there are natural kinds in the world individuated by properties existing independently of us, and that our schemes of categorization in science track these natural kinds. Here there seems a clear endorsement of the priority assumption from traditional realism. In addition, there is no constraint that the properties that feature in the cluster be observable. For example, neither the lineages of descent nor the pathways of projection from the retina to the lateral geniculate nucleus need themselves be observable to feature in the respective HPC definitions of species and retinal ganglion cells. Lingering doubts about the realist credentials of the view should be dispelled by noting that it has traditional realism as its limiting case, one in which all of the properties in the cluster are present in all instances falling under the concept. The view is a loosening of traditional realism, not an abandonment of its realist core.

Consider now the HPC view of species more explicitly with respect to essentialism and unificationism. On the HPC conception, species are natural kinds, not individuals, with essentialism in the style of traditional realism a limiting case, rather than a definitive feature of this type of natural kind. And just as the HPC view of species is incompatible with a traditional form of essentialism, so too is it incompatible with a traditional form of the commonality assumption, according to which all members of a natural kind must share some set of intrinsic properties. But there *is* a sort of common basis for membership in any given species, which can be expressed as a finite disjunction of sets of properties (and relations). We might thus view the HPC view of species as compatible with a version of the commonality thesis that allowed such disjunctions. Likewise, because some of the criteria that define the species category may have a different level of significance in different cases – in the extreme, they may be absent altogether – simple versions of the ordering assumption are incompatible with the HPC view of natural kinds. Yet the possibility of more complicated forms of the ordering assumption would seem compatible with the HPC view, since it accommodates a unified species category amongst other (unified!) biological taxa.

The HPC view of natural kinds preserves another idea that is a part of traditional realism: that all and only members of a natural kind satisfy the corresponding definition of that kind. Anything that is a species, and only things that are species, will satisfy the HPC definition for species, and any individual that is a member of a particular species, and only such individuals, satisfy the HPC definition for a particular species. Parallel theses hold likewise for neural crest and retinal ganglion cells, and their determinate forms.

But what is it to satisfy such a definition? Thus far, I have implied that it is to possess "enough" of the properties specified in the HPC definition. Here we might suspect that the vagueness that this implies regarding (say) the delineation of the species category and membership in particular species taxa is the Achilles heel of the view, in much the way that concerns about the objectivity of Autonomy pointed to problems for the Tripartite View in the previous chapter. I want to offer two responses to this concern.

First, what counts as having "enough" of the relevant properties, as with what are the relevant properties in the first place, is an a posteriori matter determined in particular cases by those practicing the relevant science, rather than by philosophers with a penchant for crisp universality. There need be no one answer to the question of what is "enough" here, but whatever answers are given in particular cases will be responsive to the clusters of properties that one finds in the world.

Second, even once there *is* general agreement about what counts as enough, there clearly will be cases of genuine indeterminacy with respect to both the species category and membership in particular species taxa. But this seems to me to reflect the continuities one finds in the complex biological world, whether one is investigating species, neurons, or other chunks of the biological nexus. There will be genuine indeterminacy about the rank of given populations of organisms, and particular organisms may, in some cases, satisfy more than one HPC definition for particular species taxa. But the former of these indeterminacies is a function of the fact that under certain conditions and over time, varieties become species, and the descendants of a given species become members of a particular genus, while the latter reflects the process of speciation (and its indeterminacies) more directly.

7. THE INDIVIDUALITY THESIS RECONSIDERED

Insofar as the HPC view of natural kinds embraces a form of essentialism, it presents an alternative to the individuality thesis and a revival of ideas central to traditional realism. Whether it represents a better alternative to the individuality thesis will turn both on broader issues in the philosophy of science and further reflection on the nature of species in particular. Let me begin here with some of the options that remain open in light of the argument of the chapter thus far.[11]

Michael Ghiselin and David Hull have presented multiple and diverse arguments for the individuality thesis about species, some for the negative part of their view (species are not natural kinds), others for its positive part (species are individuals). For example, it has been argued that the intrinsic heterogeneity within biological populations implies that species are not natural kinds, and that their status as historical entities within evolutionary theory supports a view of them as individuals. Insofar as the former types of argument presume a two-way conceptual connection between traditionally conceived essences and natural kinds, they carry no force against the view that species are HPC natural kinds. Thus, the view I have defended undermines such negative arguments for the individuality thesis. But the HPC view of natural kinds as I have adapted it here also puts both types of arguments for the Ghiselin-Hull view in a new light, because parity of reasoning should lead one to treat both species and neuronal populations in the same way: either both as natural kinds or both as individuals. Of course, such parity considerations can always be undermined by the differences between how "species" and (for example) "retinal ganglion cells" are used within evolutionary biology and visual neuroscience. But the HPC view places the burden on those who think that there is something special about species talk that warrants a unique ontological view of species as individuals to show this.[12]

Alternatively, perhaps reflection on the neuroscientific cases and these parity considerations should lead one to extend the individuality thesis beyond the case of species to other biological categories. Interestingly, this is an idea that at least some researchers in the relevant neuroscience may be amenable to. For example, following the neuroscientist C. F. Tyner, Michael Rowe and John Stone advocate what they call a *parametric* or *polythetic* approach to the individuation of retinal ganglion cells, viewing these cells not as kinds with some type of essence but as intrinsically heterogeneous populations of cells that have their own internal coherence and duration. (Indeed, they explicitly take their cue from the modern species concept.) The problem with such a view, it seems to me, is that central to neural taxonomy is the idea of identifying categories of cells that at least different organisms in the same species instantiate, and these instances considered together do not form an individual. For example, the adrenergic cells you have and those I have considered together are not spatially bounded, occupy different temporal segments, and do not form or function as an integrated whole. Perhaps this points the way to how the positive arguments for the individuality thesis can be sharpened in light of the parity considerations introduced with respect to the negative argument for it, but I remain skeptical here.[13]

This is chiefly because, indeed like neuronal populations in the heads of different agents, most species consist of spatially (geographically) isolated groups of individual entities. One of the ways in which concentration on the "biological species concept" and the associated idea of a closed gene pool has distorted thinking about species is to conjure an image of species as being more individual-like in their physical instantiation than they in fact are. Just as most groups of individual organisms, although real enough as potential units of selection, are not themselves individuals, so too are species real enough as features of the organization of the biological world despite their lacking many of the basic properties that individual organisms have.

8. PLURALISM RECONSIDERED

As a reminder of what the pluralist holds about species, consider what Dupré says in articulating his version of pluralism:

> There is no God-given, unique way to classify the innumerable and diverse products of the evolutionary process. There are many plausible and defensible ways of doing so, and the best way of doing so will depend on both the purposes of the classification and the peculiarities of the organisms in question... Just as a particular tree might be an instance of a certain genus (say *Thuja*) and also a kind of timber (cedar) despite the fact that these kinds are only partially overlapping, so an organism might belong to both one kind defined by a genealogical taxonomy and another defined by an ecologically driven taxonomy.[14]

In introducing pluralism as the denial of either or both of two assumptions central to traditional realism – the priority and ordering assumptions – I have meant to suggest that there is some tension between pluralism and realism *punkt*. The metaphysical angst that many realists experience with pluralism concerns the extent to which one can make sense of the idea that there are incompatible but equally "natural" (that is, real) ways in which a science can taxonomize the entities in its domain. There is at least the suspicion that, to use Dupré's terms, pluralism is driven more by the "purposes of the classification" than by the "peculiarities of the organisms in question," as Dupré's own analogy suggests. Such pluralism, in rejecting the priority assumption, would move one from a realist toward a nominalist view of species.[15]

Yet the most prominent forms of pluralism about species have all labeled themselves "realist," from Dupré's "promiscuous realism" to Kitcher's "pluralistic realism." Moreover, Richard Boyd views at least Kitcher's brand of pluralism as compatible with his own articulation of the HPC view of natural kinds, suggesting a form of realism that accepts the priority assumption but rejects the ordering assumption. The idea that Boyd and Kitcher share is one we saw Mishler and Donoghue express earlier: that the various species concepts that one can derive, and thus the various orders that one can locate species within, are merely a reflection of complexities within the biological world. There are two problems with this view, one with pluralistic realism itself, the other with viewing the HPC view of natural kinds as compatible with such pluralistic realism.

As pluralists say, one can arrive at different species concepts by emphasizing either morphological, reproductive, or genealogical criteria for the species category. But it is difficult to see how the choices between these sorts of alternatives could be made independent of particular research interests and epistemic proclivities, and this calls into question the commitment to the priority assumption that, I claim, needs to be preserved from traditional realism in any successor version of realism. Perhaps pluralistic realists would themselves reject the priority assumption, although Boyd's own emphasis on what he calls the "accommodation demands" on inductive and explanatory projects in the sciences imposed by the causal structure of the world suggest the he himself accepts some version of the assumption.

Boyd's own view of the compatibility of the two views fails to capitalize on the integrationist potential of the HPC view, in my view one of its chief appeals. One of the striking features of the various definitions of the species category is that the properties that play central roles in each of them are not independent types of properties, but are causally related to one another in various ways, as we saw in section 6. These causal relationships, and the mechanisms that generate and sustain them, form the core of the HPC view of natural kinds. Since the properties that are specified in the HPC definition of a natural kind term are homeostatically related, there is a clear sense in which the HPC view is integrationist or unificationist about natural kinds. By contrast, consider the view of pluralists. For example, Kitcher says that we can think of the species concept as being a union of overlapping species concepts. It appears unified in some sense, but without a further emphasis on something that plays the metaphysical role that underlying homeostatic mechanisms play in the HPC view, the unity to the species concept on Kitcher's view remains allusive. Dupré's view, by contrast, is part of an explicitly disunified view of biological taxonomy, and scientific taxonomy more generally.[16]

Consider how the differences here manifest themselves in debate over a concrete issue, that of whether asexual clonelines form species. For the pluralist, the answer to this question will depend on which species concept one invokes, in particular, whether one appeals to criteria of interbreeding to define the species category. By contrast, on the HPC view asexual clonelines *are* species since they share in the homeostatic cluster of properties that defines the species category, even though they don't have at least one of those (relational) properties, that of interbreeding. Early attempts to articulate a notion of species that applies to asexually reproducing lineages, such as that of the ciliate biologist Tracy Sonneborn, were expressed pluralistically. But the HPC view of natural kinds actually provides the basis for bringing out the agreement between these alternatives.[17]

Likewise, consider the issue of whether there is a qualitative difference between species and other (especially higher) taxa, an issue raised in the opening section to this chapter. Again, a natural view for a pluralist to adopt here is that how one construes the relationship between species and other taxa depends on which species concept you invoke. For example, on Mayr's "biological species concept," species have a reality to them provided by their gene flow and their boundaries that higher taxa lack; alternatively, pheneticists will view species and higher taxa both as nominal kinds, since taxa rank is determined by a conventional level of overall phenetic similarity. By contrast, on the version of the HPC view of species I have defended, although the general difference between various taxa ranks will be apparent in their different HPC definitions, there will be cases where questions of the rank of particular taxa remain in question and cannot themselves be resolved by the HPC view.[18]

9. REALISM, ESSENCE, AND KIND

My chief aim in this chapter has been to critically reexamine two currently popular positions about species, the individuality thesis and species pluralism, particularly in light of reflection on two cases of neural taxonomy. I have attempted to sketch how a modification of a traditional realism about natural kinds, rather than its wholesale abandonment, provides the basis for meeting the challenge of the intrinsic heterogeneity of the biological world posed by both the case of populations of individuals, species, and that of populations of neurons. Both the individuality thesis and species pluralism seem to me to be extreme reactions to the failure of traditional realism in the biological realm, and the middle-ground position that I have advocated that draws on the HPC conception of natural kinds and realism represents a plausible alternative to both the initial failure and these reactions to it. I shall conclude this discussion with one brief further comment about the HPC view, and one about essentialism.

Since the HPC view of natural kinds is in part inspired by the Wittgensteinian idea that instances falling under a common concept bear only a family resemblance to one another, it should be no surprise that it softens the contrast between natural kinds and other categories of things in the world. On the HPC view, however, two things are special about natural kinds. First, the mechanisms that maintain any given HPC are a part of the natural world, not simply our way of thinking about or intervening in the world. Second, unlike artificial or conventional kinds, natural kinds have HPC definitions that feature only properties that exist independently of us. Thus, the reason for the intermediate character of social, moral, and political categories between natural and artificial kinds is made obvious on the HPC view.

I have suggested that the concept of a natural kind needs some broadening within a realist view of science. Although I think of the resulting view as a form of essentialism, this issue is of less significance than the requisite broadening. But if we do think of the resulting view as a form of essentialism, then the concept of an essence need not be viewed as the concept of substance came to be viewed within modern science, as unnecessary metaphysical baggage to be jettisoned. Rather, species essentialism represents an important way in which Aristotle's views of the unity to the biological world, itself cast in terms of the notion of substance, have proven to be correct.

PART THREE

GENES AND ORGANISMIC DEVELOPMENT

6

Genetic Agency

1. GENES AND GENETIC AGENCY

Genes are agents crucial to the existence of many biological processes. It has long been taken for granted that genes are *the* biological agents of heredity, and they are the principal agents within genetics and its various subdivisions – developmental, population, molecular. Yet the term "genetics" was introduced by the biologist William Bateson in 1906 as the name of the emerging field studying "the physiology of Descent, with implied bearing on the theoretical problems of the evolutionist and the systematist, and the application to the practical problems of breeders," three years before William Johannsen coined "gene" to refer to the basic unit of heredity. Thus, genetics has not always been simply the study of genes. And, as we will see in more detail in Chapter 7, some recent approaches give genes a more circumscribed role in the processes of inheritance and development than they have traditionally had. In this chapter and the next we will explore the nature of genetic agency within the various areas of genetics, and reflect on what, if anything, justifies the privileged role that the gene has traditionally played in the study of inheritance and the transmission of phenotypic traits across generations.[1]

In the last fifty years, the concept of genetic agency has become integral to much thinking in biology beyond the disciplinary confines of genetics. Other areas of the biological sciences in which genes have been viewed as paradigmatic agents include developmental biology, the study of disease, and the theory of natural selection itself. Here talk of genes as programs or blueprints for organismic development, of genes for a variety of diseases and conditions, and of selfish genes, has become commonplace. In this chapter, I shall focus on genetic agency and some of the conceptions of, and claims about, the agency of genes in the biological world that both reflect and influence the perceived ubiquity of genetic agency itself.

Some of the more striking views of what genes are and what they can do have already made their appearance in *Genes and the Agents of Life*. In Part One in discussing the smallist tendencies within biology, I introduced François Jacob's use of the metaphor of a program to articulate a view of the gene as containing "the architectural plans of the future organism." As Jacob continues,

In the chromosomes received from its parents, each egg therefore contains its entire future: the stages of its development, the shape and the properties of the living being that will emerge. The organism thus becomes the realization of a programme prescribed by its heredity.

Likewise, we have also met Richard Dawkins's view that genes are the agents of selection. On Dawkins's version of genic selection, we are "survival machines" for the propagation of genes, and "their preservation is the ultimate rationale for our existence."[2]

These are striking claims, and we will discuss them in due course. My governing interest in this chapter, however, is not so much in these perhaps dramatic and memorable characterizations of genes and genetic agency as in more mundane, largely-taken-for-granted views of genes. Genes can be transmitted and expressed; they contain information or instructions; they are "for" various phenotypic traits; they can be deleted, transcribed, and transposed; they are self-replicating. Genes play diverse causal roles in biological processes, and the description of these roles crosses back and forth between the literal and the metaphorical. What sort of thing is this biological agent, the gene, that warrants these construals?

2. THE CONCEPT OF THE GENE

One answer to this question that the philosopher and biologist Lenny Moss has defended is that the gene is not one sort of thing at all, but at least two. In his *What Genes Can't Do*, Moss elaborates on a distinction that he had drawn in earlier work between two concepts of the gene, what he calls *Gene-P* (for preformationist) and *Gene-D* (for developmental resource). When biologists talk of genes for a given phenotype, such as for blue eyes, cystic fibrosis, or breast cancer, they employ the preformationist concept of the gene. What is crucial is that the Gene-P has a (contextually limited) probability of bringing about some specific phenotypic effect, and there is no commitment to the existence of some particular physical structure that corresponds to this causal role. As Moss points out, many given traits, such as having blue eyes, are said to have a Gene-P, although they are produced by the absence, rather than the presence, of a given physical resource, and there may be many physical structures that *are* the gene for a given trait. A Gene-P is an abstract entity, one invoked to play a certain instrumental role in predicting and explaining claims about phenotypes. By contrast, when biologists putatively identify genes with particular sequences of DNA, they invoke a conception of genes as developmental resources. Such a Gene-D may have different phenotypic expressions in different contexts, but what makes it the same gene across those contexts is its molecular identity. A Gene-D is a concrete, physically localizable resource, "a transcriptional unit on a chromosome within which are contained molecular template resources."[3]

Moss's claim that these are distinct concepts of the gene is the basis for two stronger claims about genetics. First, the two concepts are often confused by biologists themselves, something reflected especially in their reliance on the textual metaphors of genes as containing information and coding for traits. Second, there "is nothing that is simultaneously both a Gene-D and a Gene-P." I want first to take up this second claim, beginning with a clarification of what Moss is saying in making it. I shall return to consider the first claim in discussing the cluster of metaphors used to describe genes and the processes in which they participate at the end of this chapter.[4]

What sort of claim is Moss making here? In particular, does he mean to make a conceptual claim of some kind, or does he mean this as an empirical claim to the effect that, as a matter of fact, no one thing in the world is an instance of both of these concepts of the gene? Both interpretations seem compatible with the context in which Moss introduces the claim:

Gene-P and Gene-D are distinctly different concepts, with distinctly different conditions of satisfaction for what it means to be a gene. They play distinctly different explanatory roles. There is nothing that is simultaneously both a Gene-D and a Gene-P. That the search for one can lead to the discovery of another does not change this fact. Finding the Gene-P for cystic fibrosis led to the identification of a Gene-D for a chloride-ion, conductance-channel template sequence. But the latter is not a gene for an organismic phenotype. Its explanatory value is not realized (and cannot be realized) in the form of an 'as if' preformationist tool for predicting phenotypes. Rather, the explanatory value of a Gene-D is realized in an analysis of developmental and physiological interactions in which the direction and priority of causal determinations are experimentally first revealed . . .[5]

The first interpretation suggests, as Moss goes on to say, that the two concepts play roles in different "explanatory games," and would provide support for the idea that the confusion he diagnoses in the use of informational and textual metaphors in describing gene action is conceptual. Such an interpretation is most plausible in cases of traits, such as blue eyes (or white flowers, another of Moss's examples), in which the trait is expressed because a protein that is normally produced is absent.

The problem with this as a general understanding of Moss's claim, however, can be brought out with a parody of it. There are two concepts of a car, call them Car-P (for production) and Car-D (for driving). The first of these concepts, Car-P, is used within the automotive industry in discussing the intricate details of the production of cars, including their composition, the nature of the production lines that lead to their creation, and the economics of car manufacturing. Car-D, by contrast, is deployed by users of cars to talk about their mundane everyday interactions with cars (usually their own), such as how fast they go, what color they are, their repair costs, and so on. These concepts have different satisfaction conditions, and play distinct explanatory roles. Thus, there is nothing that is simultaneously a Car-P and a Car-D. That one can link the concepts in various ways does not vitiate this conclusion.

This parody, along with features of Moss's discussion represented in his quotation, perhaps suggest that Moss should be viewed as making an empirical claim that goes beyond the initial claim that Gene-P and Gene-D are distinct concepts of the gene: no one thing in the world is an instance of both of these concepts. This interpretation brings out a relevant difference between the Car-P/Car-D parody and the case of Gene-P and Gene-D, for in that case one and the same thing falls under both of these concepts. It also suggests that the relationship between the functional role that genes are ascribed when conceptualized as abstract entities (as Gene-Ps) and the role they are ascribed when conceptualized as concrete entities realizing these roles (as Gene-Ds) is more complicated than such ascriptions imply because different physical entities play these two roles.

The problem that this interpretation of Moss's claim faces is that even if there are cases in which those invoking the concept of a gene shift between two concepts without noticing that there is no one thing that satisfies both, there remain examples where this does not seem to be true. For example, consider Moss's own example of talk of genes for breast cancer, which he uses to show that the distinction between Genes-P and Genes-D is orthogonal to that between Mendelian and molecular genetics. BRCA1, a Gene-P implicated in some forms of breast cancer, was localized to chromosome 17 in 1990, and then further localized to a 600 kilobase region on that chromosome in 1994. This region is also a Gene-D, a developmental resource likely involved in tumor suppression, and variations in which lead, in their full context, to breast cancer. That there are different "explanatory games" played with what is referred to by "the breast cancer gene" does not imply that something cannot be both a Gene-P and a Gene-D for breast cancer.[6]

The general view exemplified by Moss's specific views is a form of pluralism about concepts of the gene, according to which there is a plurality of gene concepts, each with its own proper domain of application. The chief motivation for recent pluralist views of the gene come not so much from the traditional divide between classical and molecular views of the gene as from developments within molecular genetics itself, which has revealed more complexity to, and more diversity amongst, the kinds of genes there are than could have been anticipated fifty years ago.[7]

Evelyn Fox Keller has identified a seeming paradox in the history of genetics related to this point: that as advances in the science increasingly undermined the fundamental idea of a gene as the basic structural and functional unit in organismal development, the appeal of this very picture of the gene has nonetheless steadily increased. The "gene for" locution is widely used. Lists of genetic diseases increase significantly every year. And gene therapy is seen as a future panacea for the undesirable and debilitating. Yet the basic idea of "one gene, one protein" posited by Beadle and Tatum in 1941 required modification in light of Jacob and Monod's introduction of regulator genes in addition to such "structural genes." The discovery of split genes that led to the distinction between introns and exons in the late 1970s, and subsequent discoveries of repeated and overlapping genes, of multiple promoter sites for gene activation, and of cytoplasmic mechanisms of gene regulation, have further challenged views of the gene that make it either a structurally or functionally unified entity, let alone both. A pluralistic view of the concept of the gene seems mandated by the many different kinds of things that "gene" has been used to denote.[8]

As in the case of species, such a pluralistic view is attractive as a way to diffuse or bypass seemingly irresolvable debates in a way that preserves the idea that the major players in the debate are right about much of what they say, even if wrong in their view that theirs is the only way of properly understanding that concept. In Part Two, I introduced the homeostatic property cluster (HPC) view of natural kinds as a monistic alternative to this kind of pluralism about species concepts, and suggested that it was likely to provide an appropriate view of a range of biological natural kinds. Applying the HPC view to genes, genes should be defined in terms of some variant of the following cluster of properties:

- physically localizable developmental resource
- found on chromosomes
- region of DNA that controls a discrete hereditary characteristic
- corresponds to a single protein or RNA
- encompasses coding and noncoding segments
- reliably copied across generations through reproduction

These properties are homeostatically clustered in the sense that their coinstantiation is not accidental but the result of underlying causal mechanisms. What molecular studies have revealed is a complexity to this cluster not previously anticipated, a complexity brought to the fore by examples that show that none of these features is a strictly essential feature of genes. What Moss and pluralists more generally are correct about is that different parts of this homeostatic cluster are given special emphasis within different areas of biological inquiry. But, as elsewhere in the biological sciences, here the HPC view provides a plausible monistic alternative to pluralism.

The characterization of the gene that has perhaps been most influential both across genetics and developmental biology and in representations of the gene beyond the biological sciences was given by Richard Dawkins more than twenty-five years ago. In the next two sections, I shall examine the concept of the selfish gene, particularly as it has been used within evolutionary biology.

3. THE SELFISH GENE AND THE COGNITIVE METAPHOR

In *The Selfish Gene*, Dawkins used the distinction between replicators, individual units that can make high fidelity copies of themselves over evolutionarily extended periods of time, and the vehicles that house such replicators, to argue for the gene as the most fundamental and important agent of selection. In Dawkins's view, genes are replicators, while organisms and groups are merely vehicles. Since for Dawkins the agents of selection must be replicators, genes are the agents of selection. And since their selection is facilitated by their acting in their interests – just as is that of organisms – genes are said to be selfish. The agents of selection are selfish genes.

Prima facie, this ascription of selfishness to genes, rather than to the organisms that contain them, is an example of the cognitive metaphor. It attributes a psychological property to a kind of biological agent, in this case, the gene, that does not literally have that property. I have said that reliance on the cognitive metaphor is widespread within the biological sciences. In Part Two, I proposed that one of its functions is (to use another metaphor) to crystallize agency. Describing biological agents as if they had psychological characteristics that they do not literally have functions to heighten our sense of their status as agents, and so provides a heuristic for viewing them as the locus of causation in particular contexts. The concept of the selfish gene has just that role, not only in Dawkins's own views of the early history of life, as we also saw in Part Two, but in the expanded agentive role that genes have come to play in the biological sciences through its deployment.

One might object that my invocation of the cognitive metaphor is both misleading and unnecessary. Misleading, for it is to misunderstand the relevant notion of selfishness. And unnecessary, for what better explains the expanded role of genetic agency are advances in our understanding of how genes operate in many more biological processes than we had formerly thought. Consider the basis for the first of these claims.

There is a literal rather than a metaphorical sense in which genes are selfish, where selfishness is not a psychological property at all but one defined instead in terms of the reproductive interests of the relevant agents. To say that genes are selfish is just to say that they act in ways that maximize their reproductive interests, where these, in turn, are explicated in terms of the number of copies they leave of themselves. Genes are described not in terms of a cognitive metaphor, but on the basis of an analogy with how organisms are viewed on the standard Darwinian view of natural selection as the primary force driving evolution. That is the basis for accepting genic selection as the primary means by which natural selection operates, and for frameshifting from the standard Darwinian view of evolution with its focus on organisms to the gene as the agent of selection.

The existence of some literal sense in which we can speak of genes as being selfish, however, is something that should be expected of any metaphor that has proven useful and had influence within the biological sciences. It provides no basis for rejecting the appeal to the cognitive metaphor. My claim here, as previously, is not that appeal to the selfish gene is merely metaphorical, or more generally that cognitive metaphors cannot be "cashed out" or earn their keep. Rather, it is that there is something in addition to whatever literal understanding there is of the metaphor. The crystallization thesis aims to explain why this metaphorical extra exists.

In any case, it is not the selfish gene itself that I want to discuss further, but an idea that Dawkins has developed in its shadow. This is the idea of the extended phenotype, and a discussion of it will return us to some of the other items in the conceptual toolkit that I introduced in Part One.

4. THE EXTENDED PHENOTYPE AND WIDE SYSTEMS

The chief idea of Dawkins's *The Extended Phenotype* is that the phenotypes that express particular genes or genetic fragments do not stop at the boundary of the organism, but extend into the world at large. Shells that are found are no less part of the phenotype of hermit crabs than are shells that are grown by other crabs, and the web morphology of a given species of spider (or even individual spiders) is as much a phenotype of that species (or individual) as are the length of its legs or the distribution of pigment on its body. Phenotypes might belong to a given individual organism even though they reach beyond the boundary of that organism's body.[9]

In championing the extended phenotype Dawkins saw himself as liberating the phenotype from the bounds of the individual organism, and with it the crucial notion of phenotypic differences between organisms within a population. In Part One, I distinguished between two research strategies, constitutive decomposition and integrative synthesis, linking these (respectively) to two forms of realization, entity-bounded and wide realization for the systems explored by each methodology. The crucial difference between the two, recall, is that entity-bounded realizations are contained entirely within the boundary of the individual who has the property being realized, while wide realizations, and so the systems that integrative synthesis explores, extend beyond that boundary. Given at least the possibility of wide systems in the biological sciences, the idea that phenotypes can be and sometimes are extended in the sense that Dawkins intends seems to me both true and important. In the next chapter, I shall take seriously the idea that either or both genetic and developmental systems can be extended in Dawkins's sense.

In this section, however, I want to proceed more slowly and probe Dawkins's own view of the relationship between the extended phenotype and the selfish gene. I shall focus on three ways in which the idea of the extended phenotype as Dawkins presents and defends it is significantly more problematic than what we might think of as the bare-bones extended phenotype. I want to say why, and in so doing propose a divorce between the bare-bones extended phenotype – the idea of the extended phenotype in itself – and Dawkins's own development of it. On Dawkins's view, genes have taken on an agentive character that exaggerates their role in natural selection. I want to present an alternative way of conceptualizing the place of genes vis-à-vis the individual, one that sifts the bare-bones extended phenotype from the whole package that Dawkins advocates.

First, Dawkins presents the extended phenotype as a natural consequence of his defense of the selfish gene. (That is one reason why what in the Preface to *The Extended Phenotype* he calls the "heart of the book," articulating the idea of the extended phenotype, is to be found in three chapters that follow ten others devoted to cleaning up misunderstandings about and objections to the idea of the selfish gene.) Since genes are, to a good approximation, the only or best replicators in the evolutionary process, they are the agents of selection, and their differential survival is what matters in evolution. They replicate via the phenotypes they express, of course, but only traditional bias leads us to think of these as strictly *bodily* or *organismic* manifestations of the gene. Thus, the extended phenotype.

By contrast, Dawkins's most forceful arguments for embracing the extended phenotype are *parity* arguments that rely only incidentally on the selfish gene view. Dawkins uses widely accepted views of what sorts of thing count as phenotypes and the relation between genes and phenotypes to argue as follows: since there is no relevant difference between these paradigms and phenotypes that extend beyond the boundary of the organism, this parity consideration offers a defense of the extended phenotype. If you are prepared to accept something that grows as part of an organism as a phenotype – a shell, perhaps – why not accept something that it acquires through its interaction with the world – another shell – as a phenotypic expression of its genes? If behaviors – such as stalking in lions – can be phenotypes, as the ethologists convinced us long ago (prior to the sociobiology of the 1970s), then why not behavior that reaches into the body and behavioral repertoires of other organisms – such as that of parasitized or otherwise manipulated hosts? In Dawkins's own words, since "we are already accustomed to phenotypic effects being attached to their genes by long and devious chains of causal connection, therefore further extensions of the concept of phenotype should not stretch our credulity."[10]

Dawkins's basic point is that there is nothing in the concept of a phenotype restricting it to the boundary of the organism, a point that stands independently of the selfish gene view. As he says in several places, he is making a "logical point" about the concept of a phenotype, and as such the point has little to do with significantly more controversial views of the agents of selection. This implies that one could augment the traditional, individual-centered view of natural selection and adaptation, the idea that the individual is *the* agent of selection (or at least, in these heady pluralistic days, *an* agent of selection) with an extended conception of the phenotype. In fact, there would seem little to bar one incorporating the extended phenotype into a pluralistic view of the agents of selection that embraced even forms of group selection.[11]

Second, and relatedly, Dawkins often talks of the "extended phenotypic effects" that replicators have, the "phenotypic effects of a gene," and of phenotypes as the "bodily manifestation of a gene." This creates the impression that Dawkins thinks of phenotypes as properties of genes (as in "the long reach of the gene"), and so obscures the point that phenotypes are, in the first instance, properties of individual organisms. Genes certainly have phenotypic effects (extended or otherwise), in the sense of playing a significant causal role in bringing about those effects, but they are not the principle entity that *has* phenotypes. They are not the chief subjects of phenotypic predication. That privilege is accorded to organisms. Phenotypes do not typically belong to genetic replicators, but to the organismic vehicles in which they are housed. Eye color, running speed, and wing shape are all phenotypes of individual organisms. But so too are the extended phenotypes of web morphology (spiders), shell choice (hermit crabs), and dam size (beavers).[12]

If this is correct, then organisms are presupposed by the extended phenotype view in that they are the entities to which these phenotypes are ascribed. This means that organisms are not simply the means by which genes are packaged and propagated through generations. Rather, they are central to making sense of the extended phenotype. The view of individual organisms as mere vehicles fails to do justice to the overwhelmingly nonrandom distribution of the bearers of extended phenotypes, bearers who are the subject of generalizations about the phenotypes, extended or otherwise, that they instantiate.

Third, Dawkins contrasts organismically bounded phenotypes with those that reach into the world at large, identifying the extended phenotype with the latter. This creates the worry that extended phenotypic effects, unlike their bodily bounded kin, will be unsuited for systematic study, since the effects of genes on the world at large are infinite in number and various in strength. Call this the *dissipative concern* about the extended phenotype: that systematic study of an organism's extended phenotype is precluded because the phenotypic effects such study would require dissipate into the world at large, once we move beyond the boundary of the organism.

Consider Dawkins's idea of an "extended genetics," a supplement to conventional genetics that follows the effects of genes out into the world beyond the individual organism, with the dissipative concern in mind. If the reach of the gene is viewed as extending into the world beyond the organism, and that reach is characterized as "all the effects that [a gene] has on the world," then the organism's phenotype would include all sorts of greater and lesser effects that those genes have. Conventional population genetics is largely concerned with phenotypic variance within a population, particularly that portion due to genetic variance, and the organism serves as a clear boundary for individuating (and so measuring) phenotypic characters of study. But in an extended genetics with dissipative genetic effects this presupposition is absent, and so what variation is to range over becomes undefined. Similar problems would arise in other areas of systematic study that seem to presuppose a circumscribed conception of the phenotype, such as evolutionary taxonomy or developmental genetics. Clearly, the dissipative concern must be addressed if the extended phenotype is to be taken seriously.[13]

One way to do so is to embrace the idea that extended phenotypes are bounded by systems larger than the individual organism, wide systems. That is, by recognizing systems that include individual organisms as proper parts, as the units in which extended phenotypes are contained, we can extend the phenotype beyond the boundary of the organism without losing the focus on a bundle of phenotypic effects that could be subject to systematic study. We can make this suggestion clearer, perhaps, by considering Dawkins's own examples.

In every example that Dawkins provides – caddis fly house shape, spider web morphology, beaver dams, termite mounds, fluke parasitism in snails (and parasitism in general) – the phenotypic effects are part of some well-defined and bounded system: caddis fly plus house, spider plus web, beaver plus dam, termite(s) plus mound, parasite plus host. Thus, while phenotypes are extended in the sense of extending beyond the boundary of the individual organism to which they belong, they are not to be identified, in general, as "all the effects that [a gene] has on the world." Rather, extended phenotypes are circumscribed by individual entities larger than (and that contain) the organisms to which they belong. In short, the strategy of integrative synthesis and the corresponding notion of a wide system seem ideally positioned both to address the dissipative concern and to make sense of the idea of the extended phenotype independent of Dawkins's own selfish gene paradigm.

An alternative way to develop the notion of an extended genetics that marks a more radical departure from this organism-centered perspective is that of *community genetics*. Community genetics not only rejects the organism as a boundary for the reach of the gene, but displaces the organism altogether. As with the extended phenotype, the focus within community genetics is on the way in which the genes in one organism or population influence those in distinct organisms or populations. These are so-called indirect genetic effects, and there is particular attention within community genetics on such effects across organisms and populations that belong to different species within a larger community. Rather than simply serving as the contextual background against which changes in individual genotypes over time are measured, the genetic composition of populations itself is taken to be a causal variable that partially determines the effects that any genotype has.[14]

Some proponents of community genetics, such as Thomas Whitham, have appealed to the extended phenotype in articulating their views. They characterize the extended phenotype as "the effects of genes at levels higher than the population," however, and are primarily interested in how intraspecific genetic variation affects community organization and ecosystem dynamics. Two related features distinguish the community genetics perspective from that of the extended phenotype, both as Dawkins introduced it as an extension of the selfish gene, and in the bare-bones guise that I have extracted from Dawkins's views.[15]

First, community genetics focuses on particular "phenotypic effects" that are properties of agents larger than the individual, such as the community, in contrast to Dawkins's focus on genetic agency and mine on organismic agency. While many within community genetics have taken higher-level entities, such as communities, primarily to be the products of genetic and individual agency, there has also been research on the extended phenotypic effects that such large-scale agents themselves have. For example, there are predictable effects of the genetics of *Eucalyptus* populations on the composition and diversity of the insect communities that reside in them, such that there can be heritable differences in fitness between whole tree-insect communities.[16]

Second, on the community genetics approach, there is a two-way causal relationship between individual alleles and the genetic structure of the relevant population, rather than the latter of these simply serving as the context for the former. Thus, within community genetics the notions of community *selection* and *heritability* are well defined, and the research program fits naturally with the experimental and naturalistic studies of group and multilevel selection. As Michael Wade says in his commentary on community genetics, "[w]henever the environment contains genes, as in ecological communities, context itself can evolve."[17]

The interpretation that I have offered of both the extended phenotype and of the idea of a community genetics fit with the overarching externalist vision of biology that I have been developing thus far. In the next chapter, I shall extend this vision, and push this challenge to individualistic views of genetics further, by turning from evolutionary to developmental aspects of genetics and to development more generally. To conclude the present chapter, I return to consider the cluster of textual and computational metaphors often used to characterize genetic agency with which we began.

5. GENETIC PROGRAMS, CODES, AND BLUEPRINTS

In the introduction, I recounted François Jacob's view of the importance of the notion of a genetic program for the study of heredity. Jacob included this metaphor in the final sentence to his paper reviewing the influential operon model that he and Jacques Monod had articulated in 1961:

The discovery of regulator and operator genes, and of repressive regulation of the activity of structural genes, reveals that the genome contains not only a series of blue-prints, but a coordinated program of protein synthesis and the means of controlling its execution.

The notion of a genetic program is now ubiquitous within genetics and development. Just as the appearance of this metaphor in the 1961 paper was an important conduit for its spread into mainstream genetics, so too has Jacob's masterful history of the study of heredity, *The Logic of Life*, been an influential channel for its transmission to a broader scientific community.[18]

Evelyn Fox Keller has argued that crucial to the success of this metaphor within biology was the appropriation of the authority commanded by influential figures within cybernetics in the 1940s and 1950s, particularly Norbert Wiener and Erwin Schrödinger. The appropriation provided biologists and biochemists with a way to develop the notion of *gene action* in an empirically fruitful manner. As Keller has pointed out elsewhere, in the development of genetics in the first half of the twentieth century, "one of the great virtues of the discourse of gene action was that it permitted geneticists to pursue their research programs so productively, and for so long, without even a glimmer . . . of *how* genes act." The second half of that century did reveal much about "how genes act," and the role of the programming metaphor in those revelations requires some discussion.[19]

Part of my interest here is in the web of metaphors involved in this notion of a genetic program and the role of that web both in leading to progress in genetics in the second half of the twentieth century and in the increasingly hegemonic role that genetics has come to play within biology. Genes are strands of DNA that are said to individually code for polypeptide chains, that group-wise code for organismic character traits, and that collectively, as the genome, are said to code for the entire organism. They are transcribed into RNA, which is in turn read, edited, and proofed by various bits of cellular machinery, themselves genetically encoded. And they contain instructions that serve as blueprints or programs for ontogenetic development.

Early in the chapter, I noted Lenny Moss's claim that the coding and textual metaphors for gene action were the result of a confusion between two distinct gene concepts. Even though I have expressed some general doubts about Moss's claims about the significance of the distinction between what he calls Gene-P and Gene-D, I think that Moss is correct in calling attention to the ways in which this web of metaphors has made genes more important biological agents than they actually are. As Moss says,

Genes, like oligosaccharides, are molecular, but unlike oligosaccharides they are also conceived of as information, blueprints, books, recipes, programs, instructions, and further as active causal agents, as that which is responsible for putting the information to use as the program that runs itself.[20]

To be sure, these metaphors have advanced both genetics and its status within biology. But they have done so by facilitating simplifications of ontogenetic development that make genes appear more central to that process than they in fact are. Much like the computational metaphor within the cognitive sciences, these related (indeed, in some ways, derivative) *encoding metaphors* have capitalized on preexisting smallist biases, and have done so by localizing a kind of agency where no such agency exists.[21]

In the cognitive sciences, mental representation has been conceptualized as a form of encoding, whereby to have a mental representation, M, is to encode information about some object, property, event, or states of affairs, m. Encoding views of mental representation facilitate an individualistic view of mental representation. This is because they suggest that an investigation of the relationship between parts of an internal mental code be the focus of the study of cognition, with relationships between cognizer and environment being effectively "screened off" by the status of mental representations as codes for worldly information.

On the alternative, externalist view of mental representation that I have developed elsewhere, representation is *exploitative* in that it involves the interactive appropriation of preexisting formal and informational structures in the world, and the organism's relationship to those structures. On the exploitative view of representation, crucial to mental representation is the cognizer's active exploration of its environment through the deployment of its bodily resources, including bodily movement through the environment, eye movements, and the manipulation of external symbols, such as those, in the human case, of spoken and written language. Consider a noncognitive example that illustrates the general difference between these two views of representation.

An odometer keeps track of how many miles a car has traveled. It does so by recording the number of wheel rotations and being built to display a number proportional to this number. One way in which it could do this would be for the assumption that 1 rotation = x meters to be part of its calculational machinery. If it were built this way, then it would plug the value of its recording into an equation representing this assumption, and compute the result. Another way of achieving the end would be to be built simply to record x meters for every rotation, thus exploiting the fact that 1 rotation = x meters. In the first case, it encodes a representational assumption, and uses this to compute its output. In the second, it contains no such encoding but instead uses an existing relationship between its structure and the structure of the world, in much the way that a polar planimeter measures the area of closed spaces of arbitrary shapes without doing any representation crunching. Note that, however distance traveled is measured, if an odometer finds itself in an environment in which the relationship between rotations to distance traveled is adjusted – larger wheels, or being driven on a treadmill – it will not function as it is supposed to, and will misrepresent that distance.[22]

This latter form of representation, exploitative representation, is an effective form of representation when there is a constant, reliable, causal, or informational relationship between what a device does and how the world is. Thus, rather than encode the structure of the world and then manipulate those encodings, "smart mechanisms" can exploit that constancy. As the odometer example suggests, the encoding view also presupposes some mind-world constancy, but this is presumed only for "input" representations to start the computational process on the right track. Exploitative representation makes a deeper use of mind-world constancies.

Part of my point in introducing the exploitative view of representation has been to break the grip that encoding views of mental representation have had on the imagination of those in the philosophy of mind and cognitive science, a grip at least loosened by connectionist and dynamic approaches to cognition. Part of the point, though, is to suggest that much of the representation that cognizers like us engage in is exploitative, rather than a form of encoding. Encoding views suggest the sort of asymmetry between mental representations and other sorts of cognitive resources that individuals can avail themselves of that would underwrite individualism.

The same is true of encoding views within genetics and development. We can see this by examining the role of genes in biological development, and the interaction between the encoding metaphor and the view of genes as the primary agents for the ontogenetic development of the organism. The encoding metaphor has provided the conceptual resources for explaining facts about cellular activity (reading, translation, editing), inheritance (the transmission of information), and organismal development (program, blueprint, source code). Reliance on the metaphor has led to empirical success in biology because of an approximate, partial isomorphism between these resources and the biological processes they purport to describe. Yet one ancillary effect of this emphasis on genes, gene action, and genetic coding has been a relative neglect of the full details of both normal and abnormal ontogenetic development.

The view of organismic development as principally the interpretation of information contained in genetic codes is impoverished, and essentially omits much of the developmental detail. As Sydney Brenner said of the operon model's view that development was principally a matter of switching genes on and off: "The paradigm does not tell us how to make a mouse but only how to make a switch. The real answer must surely be in the detail." But these metaphors of genes-as-switches and genes-as-codes themselves are not fully to blame, for genes themselves play a limited causal role in the developmental economy of the organism. As proponents of developmental systems theory have reminded us, sequences of DNA in themselves are relatively inert, and there is no sense in which organisms are simply computed from the DNA that they contain. Much like the metaphor of the selfish gene, whether viewed as anchored in the cognitive metaphor or in an organismic analogy, the computational or encoding metaphor in genetics places an explanatory burden on genes that they likely cannot bear.[23]

This is a contentious claim, and debate over the significance of the textual and informational metaphors used to characterize genetic agency continues. In the next chapter, I will focus on just one aspect to this debate, one concerning the role of context in genetic coding, before going on to explore positive alternatives to the standard gene-centered view of ontogeny. Some of these are individualistic, others externalist. They include not only developmental systems theory, but more generally recent work integrating evolutionary and developmental biology, and explorations of the significance of both the extended phenotype and of the notion of niche construction.

7

Conceptualizing Development

1. GENES, DEVELOPMENT, AND INDIVIDUALISM

I have already said that genes are crucial biological agents, and we have seen some classic expressions of the idea that they are particularly important agents for ontogenetic development. And I have suggested, along with Moss, that the computational and informational metaphors used to describe genes have inflated the role of genetic agency in development. They do so in part by facilitating a conception of genes as makers of their own destiny, a conception exemplified in one of the earliest uses of those metaphors in the physicist Erwin Schrödinger's influential *What is Life?* Genes, he claimed in a well-known passage, "are law-code and executive power – or, to use another simile, they are architect's plans and builder's craft – in one." While Moss views the reliance on such metaphors as a confusion intimately tied to the melding of two distinct concepts of the gene, his Gene-P and Gene-D, I think that we should look elsewhere in trying to understand the limits to the computational and informational metaphors in genetics, particularly within developmental genetics.[1]

Genes are sequences of DNA that serve as templates for the production of amino acids. These amino acids constitute the proteins that are the basic building blocks of biological structures and processes. By virtue of this and the correspondence between certain nucleotide triplets and specific amino acids, the former are said to code for the latter. This much talk of genetic coding is relatively uncontroversial: genes code for protein synthesis. More controversial is the sense in which there is genetic encoding for much else, most notably phenotypes or even whole organisms. But prior to this extension of the encoding metaphor is a question about the metaphysics of even the encoding of protein synthesis in strands of DNA, a question that brings us back to the debate between individualists and externalists.

In Part One, I distinguished three ways in which this debate can be formulated within the biological sciences. The first simply asks whether the individual organism serves as a boundary of some kind that constrains the corresponding biological science, such that we need not look beyond that boundary in conceptualizing the natural kinds and properties in that science and the explanations they feature in. Individualists, recall, accept this constraint, while externalists reject it. Prima facie, genetics and developmental biology are individualistic in this sense, for they concern biological agents embedded deep within the organism. The processes they are active in operate fully within that milieu. I shall accept this view of genetics and developmental biology for now, but we will reconsider this in discussing developmental systems theory later in the chapter.

The second way to construe what is at issue between individualists and externalists in biology invokes the notion of supervenience and represents a more precise way of expressing the nature of the constraint that individualists accept and externalists reject. On this construal, individualism is the view that biological properties supervene on the intrinsic, physical properties of individual organisms. Since supervenience is a relation of metaphysical determination, individualism entails that two organisms can differ in their biological properties only if they differ in at least some of their intrinsic, physical properties. I have also expressed this form of individualism in terms of the notion of an *entity-bounded* realization: biological properties have entity-bounded, rather than wide, realizations. If individualism so construed is accepted, then what realizes an organism's biological properties is fully contained within the boundary of that organism. Externalists, by contrast, view many biological properties as having wide realizations, as being physically instantiated in part by what lies beyond the boundary of the organism with those properties. Again, genetics and developmental biology would appear to be individualistic in this sense, even if other areas of the biological sciences are not.

The third form of individualism is a generalization of the second. It says that the biological properties of any individual entity supervene on its intrinsic, physical properties. This can be expressed more informally as the view that taxonomy in the biological sciences individuates by causal powers: any entity's properties are determined by the causal powers of that entity. It should be clear that this is a stronger form of individualism than the second, despite the fact that the two are identical when the individual is the organism, precisely because it applies to the full range of biological agents, from genes and cells through to demes and communities. This form of globalized individualism implies that all of the properties of genetic and other developmental agents supervene on *their* intrinsic, physical properties. Their realizations are entity bounded in the strong sense that they are physically instantiated entirely by what lies within the boundary of genes or other developmental agents.

The question that I want to focus on initially in this chapter is whether genetic codes are individualistic in this third sense. Are they metaphysically determined by the causal powers or the intrinsic, physical properties of sequences of DNA? Many of the more striking claims that deploy talk of genetic coding suggest an individualistic view of genetic coding. Recall that, according to Jacob, an organism's entire future is contained within the chromosomal program, and on Schrödinger's view they are both "law-code and executive power." But this strong form of individualism is also manifest in less grandiose claims. The "Central Dogma" of molecular biology, articulated by Francis Crick in 1958, says that information begins in DNA and flows (via RNA) to proteins, but not vice-versa. Along with the companion principle that Crick articulated, his sequence hypothesis, this presupposes a conception of protein synthesis as information flow, and information can flow from A to B only if it is already in A to begin with. The sequence hypothesis says that the "specificity of a piece of nucleic acid is expressed solely by the sequence of its bases, and that this sequence is a (simple) code for the amino acid sequence of a particular protein." For example, the property of coding for the amino acid lysine is determined by the nucleotide triplet sequence AAA (as well as by AAG), which corresponds to a sequence of three adenine molecules (or two adenine molecules followed by a molecule of guanine).[2]

Those viewing nucleic acid or nucleotide sequences, in and of themselves, as constituting a code for the production of proteins, must address the following problem. Since only about 1.5% of the human genome codes for proteins, such nucleotide sequences code for proteins only in certain contexts. Sequences of nucleic acids that code for proteins are *exons*, but even apart from noncoding regions (*introns*), there are many DNA sequences that perform regulatory functions. There are promotor, terminator, leader, activator, repressor, operator, and enhancer sequences. Of such regulatory sequences, the historian of biology Evelyn Fox Keller notes:

Many of these may be considered genes in the sense that they provide templates for a gene product, but others do not 'make' anything: their function is merely to provide specific sites at which other proteins of the right kind and in the right configuration may attach themselves (bind) to the DNA.

It is not simply that more than a particular sequence of DNA is required to causally generate a protein – something disputed by no one – but that something more is required for that sequence even to be a code for a protein. Just as, upon reflection, we recognize that the fitness of an organism is not an intrinsic property of that organism, but one which is determined in part by the nature of the environment beyond that organism's boundary, so too it begins to look as though the protein-coding properties of sequences of DNA are likewise in part determined by something beyond the boundary of those sequences themselves. In short, an externalist view of protein coding (let alone of phenotypic or organismic coding), whereby part of what realizes those properties is external to the entity that has the properties, seems plausible in light of the contextual nature of that encoding.[3]

This poses at least a challenge to an individualistic understanding of the encoding metaphor, one that requires a response that explains the role of context in genetic taxonomies. Rather than develop this objection more fully, or shadow box through a possible response that could be made on behalf of the individualist here, in order to outline the contours of some of the larger issues in play I want to consider a response of this sort that has been made in a related context. Here we move from biochemistry to developmental genetics proper. I hope to show that there are quite general reasons for thinking that neither field is as individualistic as it initially appears. My focus will be on the impressive and far-reaching work on Hox genes.

2. INDIVIDUALISM WITHIN DEVELOPMENTAL GENETICS

There has been much excitement within recent developmental biology about Hox genes. Indeed, some strong claims have been made about what their discovery tells us about development and genetics. Hox genes are high-level regulatory genes that control the development of broad features of the body plan of eukaryotes. Moreover, these genes seem to have been highly conserved over evolutionary time. They are responsible in particular for regulating the formation of the anterior-posterior axis in both flies and in mice, with structurally similar genes in both of these taxa sharing also the order in which they occur on the chromosome and their expression patterns. Hox genes that are very similar in base-pair structure, such as *Hox-1.6* (in *Musculus*) and *lab* (in *Drosophila*), are considered homologues that program a developmental cascade in the ontogeny of body plans, respectively, in axially symmetrical vertebrates and arthropods.

Part of the excitement about Hox genes is that they are thought by many to provide a key to understanding organismic development, being "master control genes" whose understanding reveals just "how much of the developmental program is written into our genes." Although we can trace the research lineage leading to the excitement about Hox genes back to William Bateson's original exploration of homeotic variation and the subsequent focus on homeotic mutations in Thomas Hunt Morgan's famous fly room at Columbia University in the early twentieth century, a series of breakthroughs in the laboratory of Walter Gehring in the early 1980s heightened interest in this view of development. Central amongst these was the discovery of the homeobox in *Drosophila*, a 180 base-pair sequence of DNA that was highly conserved across different homeotic genes. When this sequence was found not only in other insects but also in vertebrates, including in the model organisms the frog (*Xenopus laevis*) and the mouse (*Mus musculus*), it appeared that homeotic genes were likely to play a significant role in programming development across much of the multicellular, eukaryotic world.[4]

I have implied that genetics is a field with an individualistic edge to its taxonomic orientation and a smallist feel to its methodological surface. From the early formulation of "one gene, one protein" to the contemporary identification of genes with sequences of base pairs, manifest particularly in the Human Genome Project, geneticists have often both bracketed the beyond-the-organism world in their taxonomies and sought to understand their basic unit, the gene, solely or primarily in terms of its constituent structures. The philosopher of biology Denis Walsh has defended this prima facie plausible view of developmental genetics by drawing on the burgeoning literature on Hox genes, arguing that the field has "earned its individualist stripes." Walsh uses the case of Hox genes to support a position that he considers a variant of traditional individualism, "alternative individualism." As Walsh's views of the taxonomy of Hox genes are developed explicitly as part of a debate between individualists and externalists, discussion of them provides an opportunity to show the significance of that debate for some of the broader issues that it raises about biological natural kinds and properties.[5]

As its name suggests, alternative individualism is a modification of standard individualism. The modification that Walsh proposes aims to allow explicitly for the role that context plays in scientific taxonomy. Taxonomy is not simply by causal powers, but by causal powers *relative to a context*. Walsh takes the individuation of Hox genes to be a case where context plays a role in an individualistic scheme of individuation within genetics and developmental biology. He views it as illustrating an individualistic position that embraces relational properties and that has much wider significance for the biological sciences.

In the next section, I concentrate on the view of Hox genes and developmental genetics that Walsh has advocated, a view that seems a natural way to understand the field, especially in light of the discovery of Hox genes. Despite the weight of tradition, scientific and philosophical, I shall argue that these views of Hox genes, and of developmental genetics more generally, are mistaken.

3. A CLOSER LOOK AT THE CASE OF HOX GENES[6]

I begin with some more information about Hox genes and their labeling. The earliest work on homeotic gene complexes was on *Drosophila*, where researchers identified the HOM-C gene complex on chromosome 3, containing at least eight identifiable genes. In their $3'$ to $5'$ order, these are *lab, pb, Dfd Scr, Antp, Ubx, AbdA*, and *AbdB*, with their expression pattern along the anterior-posterior axis of the developing fly embryo matching this physical ordering. These genes are classed as *HOM-C genes* because they are located on the same chromosome and function as a unit to specify the broad features of the anterior-posterior axis. In mammals, there are four copies of genes homologous to those within the HOM-C complex, with the suffixes "a" to "d" distinguishing these copies from one another. In mice, homeotic genes are called "Hox" genes; in humans they are labeled "HOX" genes. The anterior-posterior order of distinct genes in mammals is reflected in numbered postscripts, from 1–13, but not all copies have exactly the same genes. *HoxC*, for example, is missing c-1, c-2, and c-7; *HoxD* is missing d-2, d-5, d-6, and d-7. Moreover, since there are eight HOM-C genes, but thirteen distinct Hox genes, there are Hox genes, such as *Hox3*, that do not have a homologue in *Drosophila*. Genes with the same number across the four copies are called *paralogous* genes. Thus, "*Hoxa-4*" refers to the fourth gene on the *a* copy of the mouse

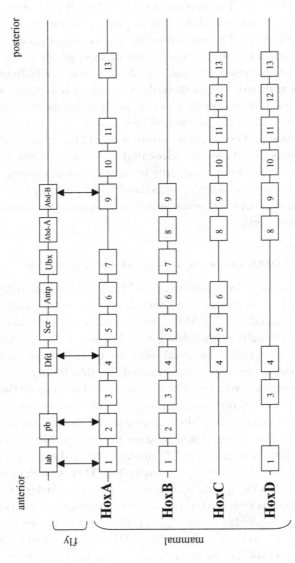

FIGURE 7.1. Hox genes in the fly and mouse (Modified from Figure 21.80 of Bruce Alberts et al., *Molecular Biology of the Cell* (New York: Garland, 1994), 3rd edition, page 1104.)

144

homologue, and it, along with *Hoxb-4*, *Hoxc-4*, and *Hoxd-4*, is a homologue to *Dfd* in *Drosophila*; the three copies of *Hox1* (given the absence of *Hoxc-1*) are homologous to *lab*. Figure 7.1 provides a perhaps more perspicuous representation of these facts about HOM and Hox genes.

One prima facie reason to think that the taxonomy of Hox genes is individualistic is that the genes seem to be identified and differentiated in terms of their intrinsic, physical properties, and thus their causal powers. *Hox1* (in mice) and *lab* (in flies) are homologues – the "same genes" in different species – because they are structurally, and hence causally, similar. The four paralogues of *Hox4* are homologous to *Dfd* because of the base-pair structure shared by these two homeobox genes. This is why they also share functions, and thus why their experimental manipulation in arthropods may reveal truths about how they operate in mammals.

Such a view, however, is incomplete in a way that calls into question the standard individualist view. Walsh himself makes this point in motivating his shift to alternative individualism. Walsh uses the fact that homologous genes, such as *Hox1* and *lab* are still classed as distinct genes, despite the similarity in their causal powers, to argue that causal powers are assessed in developmental genetics only within a context. For Walsh, vertebrates constitute one context and arthropods another. The taxonomy of homeobox genes is by causal powers within each of these contexts. That is why similarity of causal powers *across* contexts does not entail that there is no taxonomic difference between homologous genes. But there are three problems that alternative individualism itself faces. All three are related to this original problem with the standard individualist view that the taxonomy of homeobox genes is "by causal powers."[7]

The first of these turns on what Walsh means by a "context." I have just implied that for Walsh, "within a context" simply means "within an organism type" (or perhaps "within a taxon"), which is what allows alternative individualism to accommodate differential homeobox taxonomy (for example, *Hox1* versus *lab*) across taxa. The fact that homeobox gene taxonomy varies "across contexts," so construed, is also what makes alternative individualism appear compatible with versions of externalism that emphasize the importance of an entity's relational properties for its taxonomy. However, since all other aspects of the gene's contexts are irrelevant to their differential taxonomy (as *Hox1* or as *lab* genes), organismal or taxon location does not form part of the context of the gene but exhausts it: organism types or taxa are contexts. For this reason it is not really the gene's context per se that, in part, determines its taxonomy, but just the relational property of being located within a given organism type. I shall return to this point in a moment.

A second sort of problem for alternative individualism concerns not why *Hox1* and *lab* are different genes but why they are homologous. Since the two occur "in different contexts," alternative individualism is in no position to explain why they are regarded, respectively, as the *Drosophila* and *Musculus* version of *the same gene*, that is, as homologues. Here standard individualism, in looking to causal powers across contexts, is in a better position, but still it has a problem in accounting for homology, one identified, in effect, by Walsh in pointing to the need to transcend the standard individualist view. While *Hox1* and *lab* share a range of base pairs, they are not strictly identical in structure; while they have similar functions, these are not strictly identical either; and while they occur in sequences of genes that are uncannily similar, these sequences are not strictly identical. An appeal solely to an entity's intrinsic causal powers does not account for judgments of homology and paralogy. This is not because we need to relativize the appeal to causal powers within a context, as Walsh claims. Rather, it is because those judgments are sensitive to properties, such as relational properties, that aren't properly conceived as causal powers at all.

Third, alternative individualism has related problems in providing an account of what makes each of the (up to) four copies of a given Hox gene a paralogue of the other two or three within a given organism type. To parallel the response to the case of transtaxa homology, the proponent of alternative individualism should claim that despite the similarity in their causal powers, these genes (for example, *Hox4* genes) are not the same because each occurs "in a different context." Here this means not "in a different organism type," as we saw was the case for the homologous genes *Hox4* and *Dfd*, but "on a different chromosome." If this is correct, then it adds further support to my claim that Walsh's appeal to context is really an appeal to particular relational properties, rather than context per se. Yet what makes all four *Hox4* genes paralogues? Here again the standard individualist is at least in a position to point to the similarities in their causal powers. But externalists are in a better position in being able to point to these and the range of relational properties that they share. Let me explain by returning to a general contrast between individualism and externalism.

In distinguishing externalism and individualism in Part One, I said that externalists were more likely than individualists to hold that scientific taxonomies and explanations are sensitive to a wide range of factors. Externalism is accompanied by a more pluralistic view of scientific taxonomy, one that can allow some place for the causal powers of individuals but which also sees scientific (and so psychological) taxonomy in many cases as being determined by an entity's relational and even historical properties. It is for this reason that many externalists do not simply reject individualistic constraints about both cognition and biology, but are skeptical about any substantive constraints that putatively apply to large tracts of scientific inquiry.

Return to homeobox taxonomy. Externalism, in contrast to both standard and alternative individualism, can appeal to a *cluster* of intrinsic and relational features of homologous and paralogous genes, including the proteins they produce, base-pair structures, relative location on their chromosomes, place in the corresponding gene complexes, and phenotypic functions. This allows externalism – unlike standard individualism – to say why homeobox taxonomy differs even when homeobox genes are homologous or paralogous, and – unlike alternative individualism – to say why these genes are still taxonomized as the same genes. In each case, different subsets of the intrinsic and relational properties determine a sense in which these genes are the same and a sense in which they are different. The weakness of standard individualism is that it must ignore the subset of these properties that are relational (and thus not "causal powers"). The weakness of alternative individualism is that it has nothing to say about homologues across taxa (in "different contexts").

The example of homeobox genes highlights both what is wrong with standard individualism and also (and subsequently) the weakness of Walsh's alternative individualism as a modification of that view. As we have seen, the complexity to the taxonomic schemes in use in developmental genetics allows for judgments not only of whether two given genes are the same or different, but whether they are homologous or paralogous as well. To account for this full range of judgments one needs to appeal to more than the causal powers a given stretch of DNA has, as does standard individualism, and more than its causal powers in a context, as alternative individualism does.

What both versions of individualism share is the idea that causal powers have some privileged role in scientific taxonomy, summarized in the standard individualist slogan "same causal powers, same kinds," which is weakened by the alternative individualist to "same causal powers within a context, same kinds." The problem is not simply that entities rarely (if ever) have the same causal powers, but that what makes for a common taxonomy is a cluster of similar causal powers *together with a variety of relational properties.*

In Part One, I said that the appeal to an entity's relational properties in taxonomizing it is widespread in the biological sciences, and that
this posed a general problem for individualistic views. But why are relational taxonomies pervasive? This is because relational properties are
sometimes easier to discern; are sometimes presupposed as fixed in the
research traditions in which their taxonomies develop; and are sometimes those properties easiest to disturb or remove experimentally. To
subsume this range of relational properties under the heading of the
"context" in which the entities with the causal powers occur in order to
create an asymmetry between causal powers and relational properties,
as does alternative individualism, is to miss some of the complexity to
biological taxonomy, including genetic taxonomy. The pluralistic sensitivities of the externalist view of the biological sciences seem necessary
to do justice to some of the complexities of taxonomic and explanatory
practice in those sciences.

Thus far, I have challenged the prima facie view of Hox genes as individualistic by arguing that the taxonomy of homeobox genes is taxonomically externalist. I have also argued that even the least controversial
uses of the metaphor of genetic encoding – talk of sequences of DNA as
coding for proteins – presuppose a similar externalist view. But there is a
more radical form that externalism can take, one that attacks not simply
the strongest form of individualism but both of the weaker versions that
it can take and that I have granted as being prima facie plausible. This is
locational externalism. I shall argue, over the remainder of the chapter, that
locational externalism is a view of ontogenetic development that both co-
heres with and illuminates some recent departures from gene-centered
views of that process. These are developmental systems theory, the view
of organisms as themselves extending beyond their bodily boundary, the
rethinking of development in terms of the notion of niche construction,
and the appeal to notions of inheritance that are not simply genetic.

4. DEVELOPMENTAL SYSTEMS THEORY: FROM CRITIQUE TO VISION

Developmental systems theory (hereafter DST) has its origins as a critique of the perceived genocentrism within biology as it was shaped by
both Dawkins's conception of the selfish gene and by the rise of sociobiology. Its *locus classicus* is the psychologist Susan Oyama's *The Ontogeny of Information,* first published in 1985 and recently issued in a second edition. Until recently, nearly all of those who viewed themselves as

contributing to DST had their home or origin in disciplines outside of biology – philosophy, psychology, and science and technology studies being the chief three.

Many biologists, by contrast, have been skeptical of the significance of DST as a positive research program. In a paper discussing ways to resist genetic determinism that focuses on the contributions of Richard Lewontin to this issue, the philosopher of biology Philip Kitcher has gone so far as to suggest that "neither Lewontin's 'dialectical biology' nor the 'developmental systems theory' pioneered by Oyama offer anything that aspiring researchers can put to work." Such a pessimistic assessment of the significance of DST for the biological sciences seems to me premature, particularly in light of the number of active, concrete research programs within biology that can be understood through the lenses it has crafted.[8]

Kitcher is correct, however, in implying that the chief original contributions of DST to biology, and to developmental biology in particular, have been negative and critical in nature. Genes are not the principal developmental agents. There is no asymmetry between genes and other developmental agents that would justify that claim. Organisms are not simply the unfolding of a prespecified genetic program. There is no developmental biology in such a claim, only the implication that *developmental* biology is unnecessary. Development is not a passive process that simply happens to and within an organism. There are developmental patterns and variation across environments that point to the need to think about development as driven by a range of biological agents at different levels. Such negative claims do not themselves tell us, in concrete terms, how developmental biology ought to be conducted in accord with a DST vision of the field. But lurking within each of these critical claims are more positive views that, I think, do point to existing and future research, research that identifies a range of developmental resources that are themselves active biological agents in the developmental construction of the organism (see Table 7.1).[9]

The positive vision of the study of development within DST thus begins from the claim that DNA is one of a range of *developmental resources* used in ontogenetic development. Genetic resources are in no way unique or special: they are not the only entities inherited or transmitted across generations; they are not a unique or even prime cause of development; and they do not in fact code for phenotypic traits or organismic development. Or, to put it more properly: the sense in which each of these claims holds of genes does not allow one to single out genes from other developmental resources. What needs to be understood is the operation of the

TABLE 7.1. *Characterizing developmental systems theory*

1. Joint determination by multiple causes – every trait is produced by the interaction of many developmental resources. The gene/environment dichotomy is only one of many ways to divide up these interactants.
2. Context sensitivity and contingency – the significance of any one cause is contingent upon the state of the rest of the system.
3. Extended inheritance – an organism inherits a wide range of resources that interact to construct that organism's life cycle.
4. Development as construction – neither traits nor representations of traits are transmitted to offspring. Instead, traits are made – reconstructed – in development.
5. Distributed control – no one type of interactant controls development.
6. Evolution as construction – evolution is not a matter of organisms or populations being molded by their environments, but of organism-environment systems changing over time.

Source: Redrawn from Table 1.1 of Susan Oyama, Paul E. Griffiths, and Russell D. Gray, "Introduction: What is Developmental Systems Theory?" to their *Cycles of Contingency* (Cambridge, MA: MIT Press, 2001), page 2.

entire *developmental system*, including genes as a proper and important but not privileged part.

DST is not itself committed to the denial of anything except the strongest form of individualism. The basic unit of study for understanding ontogenetic development is the developmental system. While this encompasses much more than genes and DNA, one might think that it is still bounded by the organism, and that its crucial properties supervene on what lies within that boundary. Let us refer to this view as *narrow* DST, which takes developmental systems simply to be another example of a bodily system that is located within the organism and is taxonomically individualistic. Its individuation does not require moving beyond the boundary of the organism.

Narrow DST is continuous in important respects with the prevailing individualistic view of developmental biology. It is a position that Evelyn Fox Keller seems to defend, and that describes at least the early work of Eva Jablonka and Marion Lamb on epigenetic inheritance systems. Narrow DST constitutes one form of resistance to the smallist tendencies of classical and contemporary molecular genetics, at least as approaches to understanding development. This is because it denies that the organismic process of development can be adequately understood as a series of computational transformations performed by (or even from) basic informational units, genes. Rather, narrow DST seeks to identify the suborganismic system in which genes are one developmental resource, and understand how *it* operates. I presume that narrow DST could (even should) lead to a revision in the concept of a gene.[10]

Yet DST can also take a more radical externalist form, one that construes developmental systems as containing resources located in an individual's environment. This version of DST involves viewing developmental systems themselves as reaching out beyond the bodily envelope of the organism, and the externalism it exemplifies is *locational*, rather than merely taxonomic. Locationally wide DST not only constitutes a deeper challenge to individualistic views of development, but also demarcates research projects that depart from business as usual in the field.

5. DEVELOPMENTAL SYSTEMS AND LOCATIONAL EXTERNALISM

As we have seen, the debate between individualists and externalists has traditionally been construed as one over the taxonomy of scientific properties and kinds. Individualism about cognition has been based, in part, on the idea that it is either entailed or made plausible by a computational view of the mind. Computationalism provides prima facie support for individualism because it has been thought that computational individuation in general is individualistic in at least both of the first two senses of individualism that I have demarcated: it prescinds from commitments about the character of an agent's environment, and the states that it taxonomizes supervene on the intrinsic, physical properties of the agent with those states.

There is, as one might expect, a longer story to be told about the relationship between computationalism and individualism. But most relevant here is that one can challenge the inference from computationalism to individualism directly by arguing that computational systems themselves sometimes physically extend beyond the boundary of the individual whose mental states they realize. This is the central idea of *wide computationalism*. I have argued that there is reason to think that many of the mental states that cognitive agents like us have are likely to be parts of wide computational systems. Since these systems have a wide, rather than an entity-bounded, realization, their existence supports a form of externalism that is locational, rather than simply one that is taxonomic. I have also suggested that the same holds true once we relax computational assumptions about cognition, and that much of cognition more generally involves locationally wide cognitive systems.[11]

The reason for this recap of wide computationalism and wide psychology is that they represent ways of exploring alternatives to individualistic views across the fragile sciences more generally, one that is particularly apt when considering not just cognitive but *developmental* systems. An organism's developmental system need not be conceptualized simply as something that is physically bounded by its body. Rather, it is a causally integrated network many of whose components are housed not within the organism but beyond it. Thus, we can develop wide DST not, in the first instance, as a form of taxonomic externalism, but as a kind of locational externalism.[12]

Like locationally externalist views of cognition, wide DST represents a more radical break with prevailing individualistic orthodoxy, and it requires careful articulation. The key notions, in both versions of DST, are those of a developmental resource and a developmental system. There is a legitimate concern that these notions are insufficiently well defined within DST, particularly once we take the prospect of wide DST seriously, and Kitcher's claim that DST presents only a vague and methodologically empty vision for developmental biology needs to be addressed.

6. RESOURCES AND SYSTEMS: SOME CONCEPTUAL TIGHTENING

Paul Griffiths and Russell Gray define a developmental system as "the sum of developmental resources," where these are "objects that participate in the developmental process" and a developmental process is "a series of interactions with developmental resources which exhibits a suitably stable recurrence in the lineage." There seem to me several problems with such a characterization of these central notions.[13]

First, without an independent characterization of at least one of these three notions – developmental resources, processes, and systems – this is a small conceptual triangle in which to move. Second, this view leaves it unclear just why (and when) developmental resources constitute a *system,* and so has the danger of atomizing organisms into a series of independent resources. Third, in characterizing developmental resources (and so developmental systems) primarily in terms of recurrent *products* within a lineage, this view leaves itself vulnerable to an extremely expansive conception of a developmental system, particularly as one takes a liberal view of what "participates in the developmental process."

For example, consider a tree next to a bat cave around which bats must fly. This tree might recur intergenerationally and participate in some minimal sense in the process of individual bat development. Yet it would seem a mistake to include it, along with the cave, as a developmental resource. It is surely no part of the developmental system governing the bat's ontogenetic development. Intuitively, and unlike the cave itself, which affords a range of developmental and evolutionary opportunities for the bats as a species, the tree seems too peripheral to the evolution and development of the species to count as a developmental resource.

To address the first problem we need an independent characterization of at least one of these notions, ideally one that also addresses the second and third problems. Here is one suggestion. Developmental systems must be causally and functionally integrated chains of developmental resources, and these, individually and collectively, must play a replicable causal role in ontogeny and inheritance. This allows for developmental resources to include genes but also chromatin markers, cytoplasmic organelles, and protein gradients, all of which are parts of an organism that play a replicable causal role in ontogeny and inheritance. But to form part of a developmental system such resources must be causally and functionally integrated such that they collectively, as a whole, play that role. Isolated, incidental, or coincidental resources that are not so integrated do not form part of the relevant developmental system. Two things should be clear from this characterization.

First, there is a ready and principled way to extend the list of developmental resources from within the organismic envelope to beyond it, and so to move from narrow DST to wide DST. As with externalist psychology, here wide developmental systems will have realizations that cross the boundary between organism and environment. In principle, parental diets, behavior patterns, population structures, and environmental modifications, such as ant nests or beaver dams, none of which are locationally individualistic, can all serve as part of a developmental system. But to support any particular claim about something being a developmental resource in the relevant sense it is not sufficient simply to identify its replication or reoccurrence over generational time, or its causal contribution (however minimal) to survival and reproduction. Rather, one needs to chart the causal chain linking that putative resource to a series of other so-linked resources, such that they can plausibly be said to form an integrated developmental system. This requirement severely restricts the number and range of wide developmental systems there are.

Second, and subsequently, there is no single developmental system, any more than there is any single cognitive system, but many. Some of these will be narrow, others wide. Some of these will be genetic, others nuclear, cellular, organismal, or environmental. Consider four examples of locationally individualistic developmental resources.

Hox genes form part of a range of such systems. Their generic function, as we have seen, is to contribute to processes that construct gross features of bodily symmetry in animals. While Hox genes may have a high-level coding function within those systems, to think of them primarily as "master genes" for body plans would be a misleading simplification in this context. The chromatin marking system is a nuclear system containing proteins, methyl groups, and RNA complexes whose function is to facilitate the transcription of chromatin. The cytoskeleton of the nonnuclear part of cells contains actin fibers, microtubules, and intermediate filaments crucial for polymerization, chemical transportation, and mechanical structure. Aphids transmit their *Buchnera* bacteria to offspring cytoplasmically, and these form part of the developing digestive system of those offspring. Hox genes, methyl groups, actin fibers, and symbiotic bacteria all belong to distinct developmental systems that are located within the boundary of the organism they construct.[14]

Locationally externalist developmental resources tend to be identified more generically, and are often shared by individuals. Both of these features make the identification of the relevant developmental system less obvious than in the case of narrow developmental resources. Parental care is a generic developmental resource that plays a role in various developmental systems, but to specify these we need to fix on determinate instances of parental care. For example, many birds that hear their mother's song, both before and after hatching reproduce that song, which then comes to function in species-specific mate recognition. So this form of parental care forms part of their developing mate recognition system. The play that canines engage in with their young structures many species-specific behaviors, such as those involved in hunting and dominance hierarchies. Likewise, an ant nest or a beehive is a shared developmental resource that forms a part of the developmental system of many individual organisms in a colony of bees. That a parental behavior does not exclusively target an individual does not imply it cannot be appropriated as part of one or more developmental systems.[15]

While not all external developmental resources are shared by multiple individuals, many are. Part of the concern about locationally wide DST is a suspicion that the notion of a developmental system cannot be properly defined or demarcated in such cases.

7. RESOURCE SHARING AND LIFE'S COMPLICATIONS

This idea of a shared developmental resource and its role in characterizing the wide developmental systems that govern the ontogenetic development of individual organisms can be elaborated, and some of the mystery surrounding it dispelled, through consideration of the parallel with the resources that constitute locationally externalist cognitive systems. Part of the puzzle to address is how we can make sense of developmental systems that physically overlap, and what makes wide developmental systems belong to particular individuals. Consider these in turn.

The properties of the ambient optical array, of external storage systems, or of the distributed cognitive systems involved in navigation can be exploited by individual cognizers through their active, bodily engagement with these preexisting informational structures. These are shared cognitive resources, but their incorporation into the cognitive systems of individuals does not imply that these cognitive systems themselves physically overlap any more than does our digging at the same hole suggest that our actions physically overlap. Even though developmental resources appropriated through an organism's interactions with the world can involve the sharing of precisely the same physical body of matter, either cooperatively or competitively, the separation of organisms in space and time make such cases of physically overlapping developmental systems unlikely to be the norm. In any case, the main point here is that wide developmental systems need not physically overlap, even if such cases of physical overlap can be accommodated within the DST view.[16]

What of the attribution of wide developmental systems and processes *to individuals*, something that seems required if wide DST is to provide an account of the ontogenetic development of individual organisms? Again taking our cue from the cognitive case, we need to insist on the distinction between the bearer or subject of any given property and the system in which that property is realized. This is true whether that system is narrow or wide, but it is particularly important to emphasize in considering locationally wide systems. In the case of locationally externalist cognitive systems, we can view the properties and states that they realize as belonging to or governing the behavior of individual agents because it is within those agents that the *core* realization of the property is physically located. The same is true of locationally externalist developmental systems.

A core realization of a property is the physical arrangement of matter that is most readily identified as playing a crucial causal role in producing or sustaining that property. Core realizations are *partial* and are not themselves metaphysically sufficient for or determinative of the properties they realize. For example, we might think of the condition of the heart (for example, partially clogged chamber entrances) as a core realization of the property of having high blood pressure, in a given case, or of a lion's possession of large, sharp canines and claws as a core realization of its being an effective predator. In neither case are such realizations sufficient for possessing the corresponding property, for that requires that much else be true of the relevant system. This is one key point distinguishing core realizations from both entity-bounded and wide realizations.

If distinct individuals serve as the physical location for different core realizations, then this fact can be used to provide an answer to the question of how wide or externalist systems can be thought of as "attached to" particular individuals. In the cognitive case, we can identify any given psychological property or state as (for example) Marcia's because its core realization is contained *within Marcia*, even if the complete or total realization of the property is not. Likewise, we can identify properties of an individual's developmental system as Fred's because it is Fred, and not some other individual, who provides the location of the core realization of the property and system. Since many of the developmental resources that govern ontogenetic development are physically contained within specific individuals, and used primarily if not exclusively in *its* development, the embrace of even wide DST does not necessitate a shift in thinking of ontogenetic development as any less attached to individuals than do prevailing conceptions of development. Wide DST does, however, allow for the possibility that there will be cases where this link between development and individuals is either difficult to identify or breaks down completely: those cases where it is difficult or impossible to locate any core realizations within the boundary of the individual.

The key idea here is that both cognitive and developmental systems can form causal loops that extend beyond the boundary of the individual without dissipating into the world at large and without compromising organismal agency. Even wide developmental systems, like wide cognitive systems, are constituted by many resources that are themselves located within the boundary of organisms. When we move to cognitive or developmental systems for which this is no longer true, or when we are focused on explanatory questions about a series of locationally externalist resources, then the connection to individual agency is diminished.

For example, termite mounds are intricate physical structures that are built and maintained by organisms that belong to two phylogenetically distant taxa – by termites (of the genus *Macrotermes*, for example) and by the fungi that they cultivate within the mounds. Regulating the temperature and airflow within the mound is crucial to the survival of members of both species, and the mound is a developmental and ecological resource for each. But do termites build the mound in order to cultivate and harvest the fungus? Or has the fungus figured out an easy way to earn its keep by getting another species to build its home? There are reasons to answer both of these questions affirmatively, and further empirical details certainly reveal respects in which there is clearly termite agency, and others in which there is clearly fungus agency. Yet it is hard to shake the feeling that accepting both termite and fungal agency compromises the agency of each. In essence, this is a version of what in Part Two I called the Who's Zoomin' Who Problem, a problem that I suspect has only local resolutions. Moreover, when we turn our attention to shared developmental resources themselves – for example, the termite mound itself – and the broader causal nexus that they form a part of, these can take on a certain kind of agency themselves, perhaps even, as J. Scott Turner has argued, a kind of *biological* agency. This is likely to complicate any acceptable view of individuals, agency, and systems.[17]

A related complication is that developmental resources and organismic phenotypes are often intricately connected. What begins as a developmental resource, such as a tree that has rotted, may contribute to the expression of an extended phenotype, such as the dam that a beaver builds, given how a particular developmental system operates. And these extended phenotypes can, in turn, lead the organism to deploy further developmental resources, such as resultant water levels, to express further extended phenotypes, and so on. Just what an extended phenotype *is* will be as murky as (but no more murky than) the issue of what a phenotype more generally is. But conceptualizing phenotypes as parts of developmental systems at least improves the prognosis for an integrated view of development and evolution.

8. THE SMALLIST EMBRACE

We can bring this discussion of DST back to the view of genes as codes governing organismic development discussed at the end of the previous chapter. If genes are one (albeit important) developmental resource amongst many others, then the coding metaphor takes on the burden of distinguishing genes from other such resources, a burden that it is unlikely to discharge. The question here is not so much one of whether genes literally encode something, or whether this talk is metaphorical.

Rather, it is whether this talk justifies an asymmetry between genes and other developmental resources sufficient to underwrite an individualistic view of genetics. Part of the reason for focusing on wide DST in these sections is that even if there is an asymmetry in this respect between genetic and nongenetic developmental resources, this further inference to an individualistic conclusion about genetics and development is blocked by the existence of locationally externalist developmental systems, some of whose parts encode for other parts. In turn, this further deflates the significance of the encoding view of genetics, a conclusion that proponents of DST have argued for more directly.

Both taxonomic and locational externalism challenge not only our default individualistic views of genetics and development but the accompanying smallist orientation to these areas of biological science. Earlier in this chapter, I argued that the taxonomy of Hox genes was taxonomically externalist in all three senses of externalism distinguished there, and that even the property of coding for protein synthesis was externalist in the third of these senses. Hox genes themselves are located within organisms, but what makes them genes of the particular kinds that they are, what metaphysically determines how they are individuated or taxonomized, does not supervene on the intrinsic, physical properties of those organisms. Likewise, whether a given sequence of DNA codes for a particular protein is metaphysically determined by something beyond that sequence itself. One has to look beyond strands of DNA themselves in order to understand what genes they realize, not just epistemically but as a matter of metaphysics, particularly once one employs the encoding metaphor even minimally. Even locationally narrow developmental systems are likely to embody at least this much externalism, not least of all because genes are key components of many of the developmental processes that they utilize. The need to identify and examine the larger systems of which genes are a part – the need for integrative synthesis – suggests a limitation to smallist analyses within developmental biology.

The discussion of wide DST, where the developmental systems themselves are locationally externalist, poses a more radical challenge to the smallist embrace of constitutive decomposition. For it suggests that we are literally looking in the wrong place if we focus exclusively on the physical constituents of genes for an answer to our most general but also deepest questions about genetics and development. What are genes? How do organisms develop? It is not just the relational properties of strands of DNA, such as what kind of organism they are located in, but those of organisms themselves, that prove relevant to answering such questions.

9. EXTENDING PHYSIOLOGY, NICHE CONSTRUCTION, AND INHERITANCE

Returning (at last) to the concern expressed by Philip Kitcher, what of the practical upshot of DST? While Kitcher is right to highlight that much work from this perspective remains critical, the positive vision of developmental biology within DST fits well with at least three independently articulated research programs within biology that we met in passing in Part Two. These are J. Scott Turner's extended physiology, the niche construction approach to ecology and evolution championed by John Odling-Smee, Kevin Laland, and Marcus Feldman, and the study of epigenetic inheritance systems conducted by Eva Jablonka and Marion Lamb. All three of these research programs were developed independently of both DST and the debate between individualists and externalists. Nonetheless, they constitute concrete programs of research that provide a sense of how to explore developmental agency within the framework of both externalism and DST.[18]

Agents are loci of causal action, sources of differential action, and agency is not forfeited simply because some of the properties and states of agents have wide realizations. Recall that in Part Two we discussed Turner's view that an organism's physiology can extend beyond the boundary of its skin. This "extended physiology" – what would less elegantly be called a locationally wide physiology in the framework I have introduced – is the result of an examination of what is crucial to physiological processes, one that recognizes the intricate ways in which what we usually think of as basic bodily processes involve extrabodily resources. Organismic agency is not compromised simply by this consistent extension of physiological thinking, although I have indicated some of the complexities to the application of our ordinary notions of agency earlier in this chapter. Likewise, a conception of developmental systems as physically extending beyond the boundary of the organism does not in itself negate or reduce the agency of individual organisms, even if it means that we must tread more cautiously.

In Chapter 4, I said that there was ambivalence within Turner's own views on this point. The understanding of an extended physiology as a locationally wide system that, in some sense, maintains the individual organism as a core physiological agent contrasts with a view of physiological agency as frame shifting to larger units, what I called corporate organisms, that were themselves physically composed of organisms. On this latter view, the agency of individual organisms is reduced as they come to function as parts in a larger physiological agent, the corporate organism. But on either way of interpreting the idea of an extended physiology, the research program exemplifies a form of integrative synthesis, one directed beyond the boundary of the individual organism. Turner himself is characterized as a "physiological ecologist," and the integration of physiological and ecological perspectives on organismic functioning represents one of the payoffs of this externalist vision of organismic functioning.

An important connection between Turner's research program and DST is that on both an organism's physiology is not a given, or something that simply develops maturationally within the boundary of an organism. Rather, it is something that is *constructed* by that organism. This construction is made possible by the existence of organism-world constancies or regularities that are exploited by the organism over time.

This idea of an organism actively refashioning, extending, and shaping its own functioning in the world is also central to the niche construction paradigm in ecology and evolution developed by John Odling-Smee and colleagues over the past twenty years. Proponents of the niche construction paradigm have built on Richard Lewontin's rejection of a view of organisms as proposing solutions to preexisting problems posed by its environment, a view that forms part of a standard way of conceptualizing adaptation as a unidirectional process flowing from environmental circumstances to organismic responses. As Lewontin said, "Organisms do not adapt to their environments; they construct them out of the bits and pieces of the external world."[19]

Odling-Smee, Laland, and Feldman's *Niche Construction: The Neglected Process in Evolution* provides a synthetic treatment of the idea of niche construction and its significance for the integration of ecological and evolutionary perspectives on organismic agency. They define niche construction in terms of two ways in which organisms can act:

Niche construction occurs when an organism modifies the feature-factor relationship between itself and its environment by actively changing one or more of the factors in its environment, either by physically perturbing factors at its current location in space and time, or by relocating to a different space-time address, thereby exposing itself to different factors.[20]

Common external structures found across phylogenetically distant taxa, such as nests and burrows, as well as taxa-specific structures, such as dams (beavers), webs (spiders), hives (bees), and bowers (bowerbirds) require, however, not simply an act of initial construction, but ongoing maintenance and modification that shapes up a range of the behaviors of the animals who make use of them. Animals are active shapers of their environments, but they are also active responders to the changes that they have wrought in those environments. This has broad implications for how we think of biological agency.

For example, consider another of Turner's examples illustrating his idea of an extended physiology: the male mole cricket, which constructs and modifies a burrow used primarily to attract females through singing. The burrow is constructed in a funnel shape and functions as an amplifier for the tiny chirps the cricket emits. In constructing it, the cricket emits short chirps before each modification, testing the intensity of the resulting sound. This process lasts about an hour before the cricket settles into the back of the burrow and sings away for several hours. Burrow excavation is a tuning process that involves a feedback loop passing through the organism's environment, which is actively modified through that process (see Figure 7.2). Because of this, as Turner says, the "only thing the cricket need carry around in its genes is the fairly simple behavioral program for burrow building and a sensory system capable of assessing the burrow's acoustical properties and correcting its structure as needed."[21]

Odling-Smee, Laland, and Feldman have argued that in light of the pervasive character of niche construction and the contrasts between it and natural selection, "niche construction can be nothing less than a second selective process in evolution, although a very different kind of selection process from the first selective process of natural selection." On their view, niche construction is not just of developmental but of evolutionary significance. It is a process in which there can be heritable variations in fitness, but one that makes more demands on the discriminative and agentive capacities of organisms. Since the process of niche construction is not simply replicated through genetic agency, to account for its role in evolution one must extend the notion of inheritance to encompass not just genetic but *ecological inheritance* (see Figure 7.3).[22]

As I have alluded to earlier in this chapter, Eva Jablonka and Marion Lamb are well-known for their work on forms of nongenetic inheritance, what they call *epigenetic* inheritance. In other work, Jablonka has also explored the notion of *behavioral* inheritance. A paradigm of the former involves the cellular chromatin marking system, while social learning and some forms of cultural transmission involve the intergenerational transfer of behaviors. More recently, Jablonka has also discussed the idea of *symbolic inheritance systems*, which involve the transfer not simply of behaviors but of the symbol systems that (in part) generate them. She has also related her views directly to the work on niche construction and

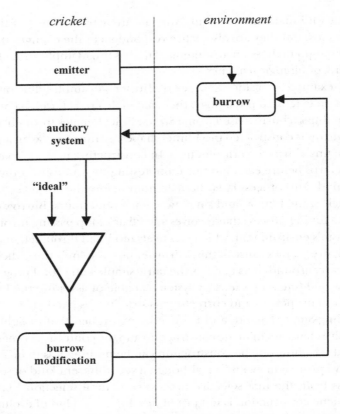

Simple model for the construction of the tuned singing burrow of the mole cricket. The cricket monitors burrow performance by how it perceives a test chirp it emits. If the perceived test chirp does not meet an ideal criterion, a round of burrow modification is initiated, which alters the energetic interaction between the sound emitter and the burrow. Note that the feedback loops extend outside the organism.

FIGURE 7.2. The extended physiology of singing cricket burrows (Redrawn from Figure 10.1 of J. Scott Turner, *The Extended Organism* [Cambridge MA: Harvard University Press, 2000], page 177.)

developmental systems theory. In the case of behavioral and symbolic inheritance systems, these developmental systems extend beyond the boundary of the individual organism, and in so doing redirect attention from what lies within the organism to how it actively deploys resources in order to achieve developmental, evolutionary, and ecological goals.[23]

The research programs that I have discussed in this section – those of extended physiology, niche construction, and pluralism about

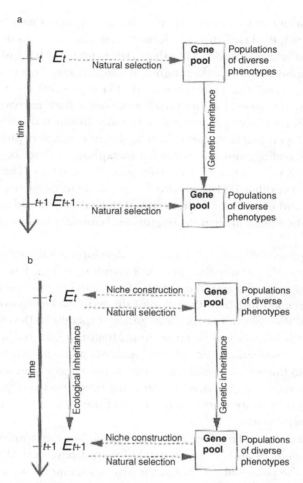

FIGURE 7.3. Ecological inheritance and the evolutionary process (Redrawn from Figure 1.3 of John Odling-Smee, Kevin Laland, and Marcus Feldman, *Niche Construction* [Princeton, NJ: Princeton University Press, 2003], page 14.)

inheritance systems – were not formed primarily as views of ontogenetic development. They offer perspectives on the biological sciences that cohere with the externalism that I have been defending. But insofar as they require one to adjust or rethink notions that are important to developmental biology – genetic and organismic agency, adaptation, evolution, and inheritance, for example – they also exemplify the sort of redirection to the study of development warranted by DST, especially wide DST.

In providing an overview to their collection of recent papers on DST, Susan Oyama, Paul Griffiths, and Russell Gray have said that many of the contributions "aim at nothing less than a root-and-branch transformation of contemporary biological thought." While that may be a sincere aim, on the view of DST that I have presented, DST represents a conception of ontogeny that is an extension of traditional views. Both narrow and wide DST constitute a challenge to some of the ways in which those views have been developed and cast, particularly to the, at times, sweeping reliance on textual, coding, and informational metaphors to describe ontogeny in terms of gene action of an especially self-contained type. The most radical alternatives that DST opens up for the study of ontogeny imply the abandonment of one or another form of individualism, and thus a tempering of the significance of the organismic boundary for developmental biology.[24]

I have proposed that the notion of a developmental system be articulated in much the way that that of a cognitive system has been conceptualized. In both cases, the notion of a system becomes more central to the corresponding field than that of the most obvious (core, partial) realizers of the system, neurons and genes, respectively. Developmental and cognitive systems can be either individualist or externalist. The emphasis on the systemic nature of development and cognition both draws attention to the ways in which the corresponding processes are causally and functionally integrated, and in so doing preempts the objection that such "systems" are arbitrary constructions of the mind, rather than parts of the physical world.

While there is a place for constitutive decomposition within both the cognitive sciences and developmental biology, I have placed more emphasis on the potential for integrative synthesis, chiefly because of the smallist neglect of the not-so-small and the way in which the individual has functioned as a boundary for scientific inquiry. In both cases, there is existing work that exemplifies the type of research relevant to going externalist. Part of the hope is to have contributed to a framework that encourages further explorations here.

PART FOUR

GROUPS AND NATURAL SELECTION

8

Groups as Agents of Selection

1. DARWIN'S LEGACY

In Part Two, I articulated what was special about organisms in the biological order of things, and in Part Three, I examined the notion of genetic agency, particularly within evolutionary and developmental biology. In the three chapters in Part Four, I turn to groups as agents of natural selection and the idea of higher-level selection more generally.

Charles Darwin is usually the person who comes to mind for most people when asked about the theory of evolution. The disciplines that the theory has spawned, the evolutionary sciences, include paleontology, systematics, and population genetics, each with their own history, methodologies, and theoretical orientations. Evolutionary approaches have also begun to make newer inroads into the fragile sciences, including inroads within psychology, medicine, epidemiology, and anthropology. Despite the diversity within the evolutionary sciences and in evolutionary approaches elsewhere in the fragile sciences, the theory of evolution by natural selection associated with Darwin represents a core theory that all share.[1]

It is common to distinguish two parts to the evolutionary views sometimes called "Darwin's theory," the first concerning evolution and the second natural selection. We can summarize these as follows:

Common Descent: new species have descended from common ancestors over evolutionary time, that is, species have evolved.
Natural Selection: the chief mechanism by which evolution from common ancestors has occurred is the process of natural selection.

Common Descent is a view about the history of species that we currently find on the planet, the view that they have descended from a common ancestor or ancestors. It implies a similar view about species that no longer exist, and those that will come to exist in the future. Thus, Common Descent claims that there is a certain pattern to the history of life: there is a *tree of life* that contains all of the species there have been, are, and will be. Since some of the species at a given time are not identical to previously existing species, Common Descent expresses a commitment to the evolution of species.

Common Descent says that the evolution of species occurs as an outcome of some process, but it doesn't imply anything very specific about that process. In particular, Common Descent itself does not imply that species are the agents of selection – that they are what natural selection operates on. Thus, while Common Descent implies that species evolve in the sense that their evolution can be seen in the pattern of species that exists over time, it does not imply that species themselves feature in the processes that produce this pattern.

By contrast, the second part of Darwin's theory, Natural Selection, is a view about the process by which the pattern in Common Descent is established. It says that the primary process or mechanism for evolutionary change is natural selection, although it too does not specify just what natural selection operates on. If Common Descent is true, then Natural Selection sketches, in general terms, why it is true. But one needs to explicate further this mechanism of evolutionary change, natural selection. In particular, we need to know what the agent or agents of selection are if we are to move beyond this sketch of a mechanism.

Darwin himself held that the primary agent of selection was the individual organism. Darwin occasionally and with caution endorsed appeals to larger individuals, such as tribes or species, as such agents, but the limitations of biological knowledge in the middle of the nineteenth century meant that Darwin did not consider smaller individuals, such as the gene, as agents of selection. In terms of the three kinds of individualism that I introduced in Part One, Darwin's view is, at least for the most part, individualistic in that it views the organism as a boundary beyond which one need not venture in exploring the mechanism of natural selection.

Stephen Jay Gould has long held this sort of view of Darwin, and has recently defended a version of this view that emphasizes Darwin's commitment to organismal selection on principled grounds. Gould claims that Darwin "reduced the locus of agency to the lowest level that the science of his day could treat in a testable and operational way – *the organism*," arguing that Darwin's revolution against the Paleyan appeal to God's benevolent intent as the cause of organismal adaptation demanded "that features of higher-level phenomenology be explained as effects of lower-level causality."[2]

Ironically, this interpretation places Darwin squarely in a general small-ist tradition of which Gould himself has been extremely critical. Gould's critique has been especially pointed with respect to the idea of the selfish gene. Claiming that "a decision to privilege the level of genes plays into the strongest of all preferences in Western science," Gould specifies this preference as "our traditions of reductionism, or the desire to explain all larger-scale phenomena by properties of the smallest constituent parti-cles." If Gould's interpretation of Darwin is correct, then Darwin himself forms part of this tradition, since organisms were taken by him to be the "smallest constituent particles" in the hierarchy of the agents of life. I shall return to consider several of the places where Darwin departs from his general reliance on organismic selection with Gould's view in mind.[3]

Following a seminal discussion of the units of selection by Richard Lewontin, natural selection is often thought of as requiring that three things hold of the entities that it acts on. On the standard Darwinian view, these entities are individual organisms, and so on that view there must be:

1. variation in a population of organisms with respect to some phe-notypic trait;
2. a correlation between this variation and the fitness levels of organ-isms within the population; and
3. the heritability of this variation across generations.

The two notions that require some brief discussion are that of a *population of organisms* and that of *fitness*.[4]

A population of organisms, as standardly conceived, is both an eco-logical and a genealogical entity. For example, the evolutionary biologist Douglas Futuyma defines a population as "a group of conspecific or-ganisms that occupy a more or less well-defined geographic region and exhibit reproductive continuity from generation to generation." So pop-ulations are intraspecific groups of organisms that are physically localized and continuous over time, including families and kin groups as well as demes. The population is physically localized and continuous over time by virtue of the physical proximity of members of a population and their temporal overlap. Indeed, we might take these facts about organisms to constitute what it is to have a population, or for a population to be a distinct kind of entity in its own right. The further claim that populations of organisms, at least as construed in the theory of natural selection, are conspecific introduces the genealogical component to the conception of a population, and privileges the species as a unique, natural unit in which microevolution takes place.[5]

An individual organism's fitness is its reproductive success, measured ultimately in terms of number of offspring that it produces. One can further distinguish between survivability and fertility as aspects of individual fitness that correspond, respectively, to the egg-adult and adult-egg halves of the life cycle, in order to attain a more fine-grained picture of where individual selection operates in particular cases, although since these have different mathematical representations, dividing fitness up in this way introduces some theoretical inelegance in modeling fitness dynamics. Individuals within the population are said to compete with one another to leave more offspring in future generations, although "competition" is used here in a "large and metaphorical sense," to use Darwin's own characterization of the idea of a struggle for existence. Whenever there is either differential survivability or fertility in a population of organisms, and finite use of resources, then there is competition, in this sense, between individuals in that population.

2. ORGANISMS AS AGENTS OF SELECTION

On this standard Darwinian view, we say that the mechanism of natural selection operates on individual organisms, or that organisms are the agents of selection. That is, individual organisms – more specifically, individual organisms with particular phenotypic characteristics – are the agents that are selected via this process of evolution by natural selection, selected because those characteristics imbue the individuals that have them with higher levels of fitness than individuals in the population that do not have them.

It has been common since the philosopher David Hull generalized Richard Dawkins's distinction between genes as replicators and organisms as vehicles to envisage natural selection as composed of two processes, replication and interaction. These may (and Hull thinks, typically do) operate at distinct levels on distinct entities. Following Hull, a replicator is "an entity that passes on its structure directly in replication" and an interactor is "an entity that directly interacts as a cohesive whole with its environment in such a way that replication is differential." Bypassing the apparent circularity in each of these definitions – at least a circularity that exists until independent characterizations are given of the processes of replication and interaction – they open the space for a two-part characterization of natural selection. With these definitions in hand, Hull defines natural selection as "a process in which the differential extinction and proliferation of interactors cause the differential perpetuation of the replicators that produced them."[6]

Following the standard Darwinian view, organisms are often taken for granted as the principal interactors in natural selection, and I take this to be central in thinking of them as the agents of selection. Although it is also common to dismiss the idea that sexually reproducing organisms are replicators, this idea deserves closer scrutiny. The reasons for rejecting organisms as replicators include appeals to facts about genetic recombination, the nonidentity of parents and offspring at the phenotypic level, and the "indirectness" of the reproduction of organisms through sexual means. But these all seem secondary and, frankly, not very good reasons, given the characterization of a replicator. Surely whether organisms are replicators turns largely on at how fine grained a level we individuate structures, for there is a clear sense in which many phenotypic structures are passed on across generations. If we individuate them relatively coarsely, many are transmitted through sexual reproduction, that is, their genetic heritability is relatively high, but that is not strictly required by the definition of a replicator. That, after all, is what makes artificial selection or selective breeding possible. But however structures are passed on, my point here is that philosophers and biologists have been too quick to dismiss the idea that organisms themselves are replicators.[7]

Elisabeth Lloyd has recently distinguished, in addition, two further issues that are sometimes built into the debate over the levels of selection, what she calls the *beneficiary* and the *manifestor of adaptation* questions. Beneficiaries of selection are those entities that end up being differentially represented in later generations as a result of natural selection. Lloyd plausibly considers species or lineages as putative beneficiaries of selection. Manifestors of adaptation are those entities that come to possess (or even lose) traits as a result of the action of natural selection. Organisms are readily conceived not only as interactors but also as beneficiaries – they are relatively easily counted, for example – and as manifestors of adaptations – their complex design was obvious before Darwin. Perhaps because of this, these distinct roles are sometimes not distinguished when considering other putative agents of selection, particularly groups, a point to which I shall return.[8]

Richard Dawkins has nicely captured one way in which the individual organism acts as a basin of attraction for evolutionary reflection:

biologists interested in functional explanation usually assume that the appropriate unit for discussion is the individual organism. To us, 'conflict' usually means conflict between organisms, each striving to maximize its own individual 'fitness.' We recognize smaller units such as cells and genes, and larger units such as populations, societies and ecosystems, but there is no doubt that the individual body, as a discrete unit of action, exerts a powerful hold over the minds of zoologists, especially those interested in the adaptive significance of animal behaviour.[9]

One of Dawkins's own chief aims, as we saw in Parts Two and Three, has been to break the hold that this image of adaptation, function, and natural selection has on biologists. He does this, recall, by making genes more focal in the range of evolutionary narratives that we tell about the biological world. In Lloyd's terms, and as she herself has argued, for Dawkins genes are replicators, beneficiaries, and manifestors of adaptation in the theory of evolution.

Breaking the grip that the individual organism exerts on evolutionary thinking has also been one of the aims of recent (and not so recent) proponents of the theory of group selection. This Return of the Group has been one of the striking developments in the theory of natural selection in the 1990s. Group selection, for decades a code word for a sloppy, mushy, and confused view of how natural selection operates in the wild, is back.

3. GENES AND GROUPS AS AGENTS OF SELECTION: A HISTORICAL SKETCH

A broad range of options is available to one interested in exploring alternatives to the individual organism as the agent of selection. Recall the genealogical and ecological hierarchies introduced in Part One. The first contains codons, genes, organisms, demes, species, and monophyletic taxa; the second, molecules, cells, organisms, populations, and communities. In this section, I shall focus on the genealogical hierarchy, particularly on the gene and the deme, returning in the final sections to say something about groups more generally as agents of selection.

The claim that the gene is the agent of selection was introduced by William Hamilton in a short paper and a pair of now classic essays on the evolution of altruistic behavior that are usually regarded as having first stated kin-selection theory. This claim also played a central role in

George C. Williams's widely-heralded *Adaptation and Natural Selection,* and has been popularized by Richard Dawkins over the last thirty years, beginning with Dawkins's *The Selfish Gene.* Genic selection has sometimes been presented as a natural extension or updating of Darwin's individual-based view of natural selection. For example, in opening his final chapter, Williams writes

Natural selection arises from a reproductive competition among individuals, and ultimately among the genes, in a Mendelian population. A gene is selected on one basis only, its average effectiveness in producing individuals able to maximize the gene's representation in future generations.

Richard Dawkins has expressed both this view of the gene as the ultimate agent of selection, and presented the selfish gene theory as a "different way of seeing, not a different theory" from the individual-centered view of natural selection.[10]

These expressions of the idea of genic selection share several features. First, each is motivated by the "problem of altruism," that is, the problem of explaining the existence of altruistic behavior, which would seem to compromise individual fitness, within the framework of the theory of natural selection. Second, each is presented in its context as an alternative to the claim that such behavior is to be explained by the existence of group selection. And third, each is taken by its proponents to be a consequence of the Modern Synthesis, the integration of Darwin and Mendel during the 1930s and 1940s, and in particular to stem from the foundations for mathematical genetics laid by Ronald Fisher, Sewall Wright, and Theodosius Dobzhansky as a part of that synthesis. Such views are individualistic in the same sense that the standard Darwinian theory of natural selection is individualistic.

Group selectionism is the view that natural selection operates not only on individuals but also on groups of individuals. As such, it departs from individualism about the agents of selection. Even if Darwin was reluctant to appeal to group selection, appeal to it he did in explaining the existence of self-sacrificial, moral behavior in human societies. In an often-quoted passage in *The Descent of Man,* Darwin says

It must not be forgotten that although a high standard of morality gives but a slight or no advantage to each individual man and his children over the other men of the same tribe, yet that an advancement in the standard of morality and an increase in the number of well-endowed men will certainly give an immense advantage to one tribe over another.[11]

While a "high standard of morality" may be of no advantage to the individual within a group, it may provide an evolutionary advantage to the group itself in its competition with other such groups. This suggests that a "high standard of morality" could evolve by group rather than by individual selection. We will be in a position to appreciate the relationship between Darwin's suggestion and contemporary views of the problem of altruism in section 4.

Group selection was often invoked by biologists from the 1930s until the 1960s. Most notable here are Alfred Emerson's contributions to the superorganismic tradition within entomology. Emerson formed part of what became known as the Chicago School of ecology whose work culminated, in many ways, in the 1949 book *Principles of Animal Ecology*. Together with his colleagues Warder Clyde Allee and Thomas Park, Emerson articulated a view of social life as harmonious and well-adapted for integration into the natural world more generally. This general view was supported by the idea that social groups could often be viewed as superorganisms, a view that Emerson defended via an extensive physiological analogy between organisms and the mechanisms that maintain and propagate their existence and the structure and functioning of colonies of social insects (*Hymenoptera* and the termites). Such groups thus come to bear a range of adaptations that are the product of natural selection and give those groups a certain level of fitness. As Emerson said, "evolution has resulted in an integrated, balanced, biological system incorporating organisms of various species and various organismic levels, in its entirety exhibiting dynamic equilibrium between its parts and with its external environment."[12]

A popular way of expressing such views was to say that individuals act "for the good of the species," or that groups act "for their own good." A passage that William Hamilton cites in the introduction to the reprint of his influential "The Genetical Evolution of Social Behaviour, I and II" gives a caricature of these views:

Insects do not live for themselves alone. Their lives are devoted to the survival of the species whose representatives they are. . . . We must now stand back and look at the insect as a member of the 'population' or 'species' to which it belongs. Indeed we have now reached the heart of the matter – the aim and purpose (so far as we can understand them) of the life of insects.[13]

In general, the idea of group-level adaptations was viewed as unproblematic, and it was invoked to explain a wide variety of biological phenomena. To again draw on Lloyd's terminology, for these group selectionists groups were not only interactors but the beneficiaries of selection and the manifestors of adaptations.

Explanations provided by the Chicago School and by the Scottish biologist V. C. Wynne-Edwards became particular targets for George C. Williams in *Adaptation and Natural Selection.* Largely through the subsequent influence of this book, group selection has received short shrift within evolutionary biology until the last decade or so. Williams introduced his own version of C. Lloyd Morgan's Canon – never invoke a higher-level mechanism when a lower-level mechanism will suffice – stating that

adaptation is a special and onerous concept that should be used only where it is really necessary. When it must be recognized, it should be attributed to no higher a level of organization than is demanded by the evidence. In explaining adaptation, one should assume the adequacy of the simplest form of natural selection, that of alternative alleles in Mendelian populations, unless the evidence clearly shows that this theory does not suffice.[14]

On one reading, the principle in this passage, what I shall call the No Higher Than Necessary Principle, is unobjectionable, even trivial: let theoretical sophistication follow "the evidence" for the need for that sophistication. But from that reading very little follows.

To derive more substantive conclusions about adaptation and natural selection, one needs to attend both to the equation between lower and simpler levels of selection, and to the presupposition that there are at least a range of cases for which an appeal to "alternative alleles in Mendelian populations" – to genic selection – is explanatorily adequate. Both assumptions are controversial – the first manifests a sort of smallist bias, while the second is challenged by the prevalence of physiological epistasis, for example. But my main point here is just that the No Higher Than Necessary Principle is subject to both readings, and its plausibility is likely heightened by failing to distinguish these.

To this principle Williams added another, and together these had considerable influence in damping enthusiasm for group selection. This second principle concerns the relationship between the level at which adaptations can legitimately be ascribed, and the level at which the mechanism of selection operates. In Williams's view, adaptation at a given level must be underwritten by a process of natural selection at that same level, and so "group-related adaptations must be attributed to the natural selection of alternative *groups* of individuals." Williams initially used this principle, matching the level of adaptation to that of selection, to question whether there were *any* group-level adaptations. This Matching Principle (as I shall call it) also played a role in his later endorsement of a limited form of group selection for group-level traits, such as optimal mutation, recombination rates, and the evolution of sex ratios. For Williams, one of the chief reasons for skepticism about what he called "biotic adaptations" was that while it was relatively easy to describe populational and other large-scale organizational structures of the biological world as if they bore adaptations, it proved significantly more difficult to specify mechanisms that operate at a corresponding level of organization. And, in accord with the Matching Principle, when there are no mechanisms at a level, there is no adaptation at that level.[15]

The Matching Principle, however, is at least as questionable as the substantive reading of the No Higher Than Necessary Principle, particularly when considered in the context in which group selection has been discussed. That the levels of selection and adaptation can come apart is presupposed by the very idea that altruism, an individual level trait, might be explained by an appeal to group selection. Hamilton and Williams questioned the need for this appeal, but not its coherence. Moreover, Williams himself was keen to show that group-level traits were nearly always to be explained by an appeal to lower-level mechanisms, such as individual selection.

Whether such considerations ultimately undermine the line of thought that Williams directed against group selection, however, seems to me doubtful. This is because of the two prongs to the individualistic critique that he offered.

On the one hand, the individualist is likely to view any putative group-level adaptation as an adaptation at some lower level. As Williams himself said, a fast-running group of zebras is nothing but a group of fast-running zebras. And presumably altruistic groups are just groups of organisms that are altruists. On the other, the individualist will question whether we are forced to appeal to group selection in any given case. The phenomenon of female-biased sex ratios is a classic case here. This was an example cited by Williams as in principle providing strong evidence for group selection if found (Williams himself thought that sex ratios were balanced). But just a year later, William Hamilton argued that female-biased sex ratios were prevalent in many species of arthropods, yet that this could be explained by an *individualistic* theory of selection of the type that he had presented in his earlier papers. The general issue of the dialectic between individualists and their group selectionist rivals will receive greater attention in the next chapter when we examine in detail one other frequently cited, putative case of group selection, that of the evolution of the myxoma virus.[16]

Group selection has been revived by the biologist David Sloan Wilson and the philosopher Elliott Sober very much in the wake of the critique delivered by Williams. (Sober's first acknowledgment in the preface to his first book in the philosophy of biology is to Williams for writing *Adaptation and Natural Selection*.) For example, Wilson and Sober share Williams's Matching Principle, which links the levels of adaptation and selection. The proponents of this "new" group selection have also self-consciously distinguished their view from "old" group selection, and have developed mathematical models of how group selection operates in specific cases. Wilson's early work had this focus, and his more recent work has generally looked to broaden the significance of this revival of group selection. Most influential, particularly within the philosophy of biology and amongst nonbiologists, has been the joint work that Sober and Wilson have produced.[17]

As I have indicated, the problem of altruism has played a pivotal role in debate over the agents of selection. It will pay to have a careful statement of the problem before us.

4. THE PROBLEM OF ALTRUISM

On the standard Darwinian view, populations of organisms evolve because the individuals in them have differential levels of fitness. As we have seen, those organisms can be said to compete with one another "in a large and metaphorical sense" for the survival of their offspring. In this same sense, organisms can be thought of as striving to maximize their fitness, that is, their own survival and ultimately the survival of their progeny. Although organisms are often thought of as striving for their own survival, those that do so to the exclusion of producing viable offspring – for example, either by producing no offspring at all or producing none that survive as fertile individuals – have a fitness of zero.

Given that the fitness of any given organism is ultimately its expected number of offspring, any individual striving to maximize its fitness will be striving to maximize this number. Thus, it will act in ways that benefit at least some others, that is, its progeny. But an individual's biological fitness places it in competition with other members of the population, and so individuals who reduce their own fitness in order to increase the fitness of others who are not progeny will reduce their representation in future generations.

Evolutionarily altruistic behavior is typically characterized as behavior that has just this property of reducing an individual's fitness while increasing the fitness of nonoffspring in that individual's group. For example, Edward O. Wilson defined altruism as "self-destructive behavior performed for the benefit of others" in the glossary of his influential *Sociobiology: The New Synthesis*. But such behavior is merely an extreme form of a more general type of behavior that gives rise to the problem of altruism.[18]

This problem arises just when a behavior contributes relatively more to the fitness of nonoffspring in the population than to the fitness of the individual engaging in the behavior, and thus that decreases the *relative* fitness of the "altruistic" individual within the population. Because individual selection will diminish the relative fitness of individuals engaging in such behaviors from one generation to the next, it will select against them. If unchecked, it will drive them to extinction in the population. It is precisely such behaviors that give rise to the problem of altruism. Thus, I suggest that these – which may or may not be "self-destructive" or "performed for the benefit of others" – be considered altruistic behaviors, that is, behaviors for which the problem of altruism arises. Behaviors that are self-sacrificial or that benefit others should be considered merely as a special case.

Since this characterization might be thought to distort the problem of altruism, or to introduce a new problem under that name, let me explain in more detail why this is a preferred way of conceiving of that problem, and not a change in topic. First, it makes clear that the crucial notion is that of an individual's relative fitness within a population, with the problem being to explain how behaviors that reduce that property of individuals could evolve or be preserved over evolutionary time. Second, it softens the traditional focus on the contrast between "selfish" and "altruistic" behaviors by making it clearer that both self-beneficial and other-costly behavior are also subject to the problem of altruism. Self-beneficial behavior is so subject just when it is accompanied by more strongly other-beneficial behavior. And other-costly behavior is subject to the problem of altruism just when other-costly behavior is accompanied by more strongly self-costly behavior. In both of these cases, the result and the explanatory problem is the same. Relative fitness decreases, and so one needs to explain how individuals with traits that decrease their relative fitness in a population can survive over evolutionary time. Third, and relatedly, this characterization of the problem of altruism makes it easier to dissociate the problem from the idea that self-sacrifice is at the heart of altruism, and the psychological caste that that notion has to it. The sacrifice involved in behaviors for which there is a problem of altruism is just that of the maximization of the number of one's viable offspring.

Given the individual as the agent of selection, the existence of altruistic behaviors, so characterized, would be a puzzle, since individuals in a population who exemplify them will be less fit than those who do not. Thus, other things being equal, such individuals will leave fewer offspring in the next generation than do their competitors. From this perspective, being altruistic is a differential handicap, like being slow relative to others in a population, where greater speed allows one either to capture more prey or to escape more readily from predators. Such fitness-reducing behaviors may be the by-product of selective processes operating on other phenotypes but could not themselves evolve by individual selection.

5. ALTRUISM BEYOND THE STANDARD DARWINIAN VIEW

The problem of altruism, then, is the conjunction of the standard Darwinian view of natural selection with the existence of evolutionary altruism. There are thus two ways to respond to the problem that could be said to represent solutions to the problem, rather than either an admission that the problem reveals the limits of the theory of natural selection (defeatism), or a denial that there is a problem at all for the standard Darwinian view to face (blind optimism).

The first is to deny the existence of evolutionary altruism. Given a range of often-cited cases – for example, sentinels in birds, caste specialization in social insects, "good Samaritan" behavior in humans – in which individuals help others or even sacrifice their lives for others – such a denial might be thought to lack prima facie credibility as a response to the problem of altruism. However, altruistic behavior is not simply helping or sacrificial behavior, but behavior that detracts from the relative fitness of the individual. So to demonstrate the existence of evolutionary altruism one cannot simply point to clear instances in which individuals help others or sacrifice themselves for the sake of others. For such behaviors might themselves be a way of maximizing individual fitness.

Indeed, this is the idea behind reciprocal altruism: individual's forego or limit their own direct reproductive opportunities in order to maximize their long-term fitness through gaining reciprocal benefits from those they benefit. Here individuals are still maximizing their own fitness, albeit indirectly. Hence, these behaviors only appear to be evolutionarily altruistic. In effect, this response plays up the role of individual fitness within evolutionary theory so that there is little or no room for evolutionary altruism. For it to solve the problem of altruism the net benefits to individuals engaged in "altruism" must be greater than the net benefits to those they help.[19]

The second is to modify the standard Darwinian view so as to posit some other unit of selection, and then show how selection operating at that level could give rise to evolutionary altruism. Thus, proponents of group selection have pointed out that although individual selection acts to decrease the representation of altruists within a population, groups of altruistic individuals may have a higher level of fitness than nonaltruistic groups. It follows that a process of group selection will act in a countervailing direction to that of individual selection, and thus altruists could survive as members of fitter groups. This version of the second response goes hand in hand with the idea that the traditional Darwinian view requires augmentation, and that there is a plurality of levels at which natural selection operates.

Darwin himself made appeals of both general sorts in his discussion of the evolution of sterile insect castes in Chapter 7 of *On the Origin of Species*. As organisms that do not themselves reproduce at all, members of such castes as worker ants and bees have a fitness of zero. Their prevalence represents an extreme form of the problem of altruism. Having noted this "one special difficulty, which at first appeared to me insuperable, and actually fatal to my whole theory," Darwin argues that this difficulty "is lessened, or, as I believe, disappears, when it is remembered that selection may be applied to the family, as well as to the individual, and may thus gain the desired end." Here Darwin appeals to groups (families) as the agents of selection, just as he later appealed to groups (tribes) to explain the evolution of morality. But alongside this claim, Darwin also introduces the idea that sterility may be a by-product of the selection of some other characters. Darwin "can see no real difficulty in any character having become correlated with the sterile condition of certain members of insect-communities," claiming that groups with a single caste may have developed "by the long-continued selection of the fertile parents which produced most neuters with the profitable modification." This is individual selection, and the explanation here parallels Darwin's account of hybrid sterility in the following chapter of the *Origin*.[20]

An alternative way to depart from the traditional Darwinian view is more radical in that it involves recasting the theory of natural selection (and thus fitness) in terms of the survival not of organisms but of the genes they contain. If genes are the agents of selection, then organisms can be altruistic if their behaviors maximize the fitness of genes that happen to be located within those organisms. Since not just progeny of a given organism but individuals related in other ways to it, such as siblings and cousins, bear a genetic relationship to that organism, altruism directed at those individuals may be a way of maximizing the fitness of that organism's genes. This is a common way of understanding Hamilton's inclusive fitness or kin selection theory. In effect, this view also denies the existence of evolutionary altruism, and thus implies that both conjuncts that constitute the problem of altruism are false.

There is an important asymmetry between genic and group selection implicit in what we have already said that can be made more explicit by posing two questions:

a. Does the traditional Darwinian view provide us with a complete or exhaustive view of evolution by natural selection?; that is, are there evolutionary phenomena that this conception of the agent of selection leaves out?
b. More radically, are the appearances here actually misleading?; that is, are there other agents that are in general better candidates for the agent of selection than the organism?

Proponents of genic selection answer "Yes" to (b) because they think that genes are better candidates than organisms for the agent of selection. In part, this is because the gene's-eye-view of evolution provides a solution to the problem of altruism. But it is also because this view is thought to have other admirable features, such as pinpointing the real underlying mechanisms governing natural selection or providing a more parsimonious framework for representing the action of selection, features captured in the No Higher Than Necessary Principle and the Matching Principle.

Proponents of group selection, by contrast, answer "No" to (a) because they think that certain phenomena (for example, altruism) require group selection. Thus, they hold that such a process must be added to individual selection to understand the complexity to the biological world. In fact, proponents of group selection are typically happy enough to embrace levels of selection smaller than the organism, such as the gene, as part of an overarching *multilevel* approach to understanding natural selection. In what follows I shall concentrate chiefly on groups as agents of selection, beginning with some further articulation of group selection.[21]

6. ARTICULATING GROUP SELECTION

There is an intuitive path to and expression of the idea of group selection that it may pay to have before us from the outset. Just as there can be the natural selection of organisms within a population for some fitness-enhancing property – running speed, wing shape, color – so too can there be the natural selection of groups within a population of groups for some fitness-enhancing property. This selection of groups is group selection, just as the selection of individuals is individual selection.

A common and almost immediate reaction by the biologically unini-tiated and initiated alike to this quick path to group selection is to view groups as more problematic entities than organisms vis-à-vis the agents through which the process of evolution by natural selection acts. There is something to such a reaction, and I shall consider it further after show-ing how to match this intuitive expression of group selection to a more precise characterization that makes it clear how group selection is a form of natural selection. Furthermore, we need to be clearer on the nature of the agentive role that groups play in this process: are they interac-tors, beneficiaries of selection, or manifestors or adaptations, or some combination of these?

The most convincing way to provide a more precise characterization of group selection is to show that there are strict analogs to Lewontin's three conditions for natural selection – (1)–(3) from the first section of this chapter – that take the group rather than the individual organism to be the relevant object of focus. These analogs are that there must be:

1.′ variation in a population of *groups* with respect to some phenotypic trait;
2.′ a correlation between this variation and the fitness levels of *groups* within the population; and
3.′ the heritability of this variation across generations.

Note that (1′)–(3′) are derived from (1)–(3) simply by substituting "groups" for "organisms" in the two places that I have italicized. The paradigm of laboratory or experimental group selection established by the geneticist Michael Wade over thirty years ago is based on the satisfac-tion of (1′)–(3′), much as the domestic breeding that Darwin appealed to in the opening chapter of the *Origin* was based on the satisfaction of (1)–(3).[22]

With individual selection, "individual organism" is left as an intuitive and undefined term, and the characterizations of what a population of organisms is, and what organismic fitness is, are relatively straightforward, as we have seen. And since organisms seem robust enough entities in their own right, this is not without some justification. The same, however, cannot be said of groups.

The close parallels between our three conditions for individual and group selection suggest what I shall call a *parasitic strategy* for articulating group selection. I propose to develop this strategy in responding to concerns about the ontological status of groups, and in clarifying the role of groups as agents of selection.

7. THE ARBITRARINESS AND EPHEMERALITY PROBLEMS

A number of the doubts about the coherence of group selection turn on concerns about the *ontological* status of groups. For example, there is the suspicion that there is no way to demarcate groups of organisms from mere sets of organisms, and so to specify just which group is the agent of selection in any particular case. This would imply that any n-tuple of individuals could be said to constitute a group, and that there is no constraint on just what natural selection might then be viewed as operating "for the good of." For example, Richard Dawkins playfully notes that "[l]ions and antelopes are both members of the class Mammalia, as are we. Should we then not expect lions to refrain from killing antelopes, 'for the good of the mammals'?." And there is the objection that groups, unlike individuals or genes, are not sufficiently permanent arrangements in the biological world to be the units on which natural selection operates over evolutionary time. Dawkins and George Williams are both well known for characterizing groups as temporary features of the biological world, not persisting over evolutionary tracts of time as, they claim, do genes. In sum, groups are *arbitrary* collections of individuals that are too *ephemeral* to be subject to natural selection.[23]

Suppose, however, that *a group is a population of organisms in just the sense invoked in (1).* That is, a group is an intraspecific, typically physically-bounded, ecologically integrated population of individuals – something like a *deme.* This is the conception of a group operant in both experimental and theoretical work on group selection, and it provides a way of extending the parasitic strategy. On this view, (1') requires only that there be a population of populations of organisms, and that that metapopulation varies with respect to some trait. That is, there is variation across the groups in the metapopulation with respect to some trait, just as there is variation across the individuals in the group with respect to some phenotypic trait in individual selection.

This trait could either be a property of individual organisms, a classic phenotype, just as in the case of individual selection, or it could be a property of the groups themselves, a *group-level trait*. Group-level traits are of two types. First, they may be *group-only traits*, possessed only by groups and not by individual organisms. Relevant examples in this context are population size, genetic variability, or migration rate, properties that individual organisms cannot have. Second, there are traits, such as cooperativeness and reproductive capacity, which can be possessed both by groups and by the individual organisms that constitute them. These latter group-level traits are what I shall call *multilevel traits*.

Thus, we relieve worries about the ontological status of groups by parasitizing the taken-for-granted notion of a population of organisms in the standard Darwinian view of natural selection. Groups so characterized are neither arbitrary nor ephemeral collections of individuals. They are not arbitrary because conspecific, physically localized grouping is a distinctive feature of the biological world. And they are not ephemeral because such groups continue to propagate themselves as generations replace one another in the group.

The power of this response to the skepticism about the ontological robustness of groups is that it utilizes the very notion of a population extant within the theory of individual selection. Thus, group selection is as robust, and as labile, as that notion. Given that populations are typically conceived as intraspecific groups of organisms, so too are the metapopulations made up of such groups. Thus, there will be an evolutionary structure to the metapopulation itself.

There are two quite distinct conceptions of a group that have been posited in the revival of the theory of group selection, that of a *trait group* and that of a *superorganism*. These are sometimes run together, and discussions that begin with one often shift to the other, typically in making an argumentative point. In the next section, I discuss these two conceptions in light of the arbitrariness and ephemerality problems, the parasitic strategy, and the distinct roles that we might ascribe to groups as agents of selection.

8. TRAIT GROUPS AND SUPERORGANISMS

David Sloan Wilson introduced the idea of a trait group as a part of his original theory of group selection, one of whose aims was to serve as a model of natural selection that was not purely individualistic. There Wilson characterized trait groups as intrademic structures, "populations enclosed in areas smaller than the boundaries of the deme," where these populations are defined relative to some specific trait. Demes are usually taken to be local populations in which mating is random, and so are the largest population analyzed by standard models of population genetics. Trait groups were initially posited as groups within a population of conspecifics (via the notion of a deme), "structured demes," as Wilson called them, thus constituting the basis for articulating the evolutionary structure to that population. But the intuitive idea of a trait group is more general, resting as it does on the notion that traits have a certain "sphere of influence" in which their effects are felt. Examples of trait groups are caterpillars of the same species of butterfly that feed on the same leaf, or vessel-inhabiting mosquitoes. Such trait groups could be subject to selection pressures different from those elsewhere in the structured deme, and thus could form the basis for group selection. Wilson used the intuitive notion to describe the evolutionary dynamics within multispecies communities.[24]

In joint work with Elliott Sober, Wilson has more recently built on this intuitive notion by characterizing a trait group as "a set of individuals that influence each other's fitness with respect to a certain trait but not the fitness of those outside the group." On both this and the earlier view, how long a group persists is irrelevant to its status as a group, as is whether the group has a relatively well-defined spatial boundary. The crucial notion, rather, is that groups contain members who interact in some evolutionarily significant way. The more recent characterization of a trait group explicitly endorses a notion of group selection that is interdemic (rather than simply intrademic), and so allows for group structures that are defined in terms of multispecies or community interactions. For example, groups of symbiotic bacteria and their hosts, such as *Wolbachia* bacteria and parasitoid wasps in the genus *Nasonia*, or the groups formed by phoretic associations between mites and carrion beetles in the genus *Nicrophorus*, constitute trait groups on this second characterization.[25]

These views of groups effectively dismiss the ephemerality problem as a nonproblem, and respond to the arbitrariness problem by appealing to the *shared fate* of members of a trait group. Members of trait groups share an evolutionary fate in that they face the tribunal of selection together, and they do so as members of that trait group. Given that, trait groups are best seen simply as interactors, and not, in addition, as either the beneficiaries of selection or the manifestors of adaptation. Temporary and spatially dispersed trait groups may well be agents of selection, but they need not themselves benefit from this agency, nor need they come to possess adaptations as a result of it, for the simple reason that they may not continue to exist over evolutionary time. That is, trait groups need not be what increase their proportional representation in the metapopulation, nor need they be what possesses complex adaptations as a result of natural selection operating on them.

Kim Sterelny has pointed to two prima facie problems with this notion of a group. The first is that the two criteria given for demarcating trait groups – that of sharing some specific trait (such as feeding on the same leaf or sharing a local environment more generally) and that of falling within a sphere of influence – do not always pick out the same groups of organisms. Organisms that, say, share a local environment may behave differently in it (including in how they interact with one another), and so may have different evolutionary fates. And organisms that are, in some sense, tied together and have a common evolutionary fate may do so by virtue of the very different properties they possess, as the examples of multispecies trait groups suggest. Since the idea of a sphere of influence underlies the explicit characterization given to trait groups in the early paper by Wilson and has been further developed in later work, I shall assume that it is the criterion that it provides for demarcating groups that is primary.[26]

The second problem, given that, is a version of the arbitrariness problem: just which organisms and species to include within any given trait group. Influence is a matter of degree, and the ecological interconnectedness that is taken seriously in Wilson's application of trait group thinking to multispecies communities raises questions about just where many trait groups begin and end. Consider Wilson's original example of a butterfly that lays its eggs on particular leaves, such that its caterpillars share a local feeding source that may differ from that of other local populations. The activities of these caterpillars influence the evolutionary fate not only of one another, but also of both other populations of the same species elsewhere and of other species, such as competitors as well as the plants bearing the leaves. The influence of caterpillars on a leaf clearly extends beyond that group, and while we might recognize it as a real group (that is, an interactor), multispecies groups that feature it seem equally real, by this criterion. Conversely, if we inspect the sphere of influence within our group of caterpillars we may find differences that suggest distinct subpopulations. I suspect that Wilson and Sober think that we should be pluralists about such groups, placing emphasis on their distinctness as trait groups by recognizing that they are defined with respect to different traits. The plausibility of this response is something I will return to in the next chapter in turning to a detailed examination of a putative actual case of group selection "in the wild."[27]

An alternative and prima facie very different way to think about groups is in terms of the concept of the superorganism. A superorganism is a group of individuals that itself functions as an individual, and whose parts, individual organisms, function as organs of that individual, each with its own function to perform. The superorganismic conception of a group is particularly pronounced in Wilson and Sober's *Behavioral and Brain Sciences* paper, where one of their central conclusions is that "higher units of the biological hierarchy can be organisms, in exactly the same sense that individuals are organisms, to the extent that they are the vehicles of selection." Wilson and Sober claim that "it is legitimate to treat social groups as organisms, to the extent that natural selection operates at the group level" and to view groups as becoming organism-like through the process of group selection. This is a conception of groups as functional unities that emerge from the evolutionary order under certain conditions, becoming like the entities, individuals, from which they emerge.[28]

As individuals, superorganisms are neither arbitrary nor ephemeral entities. But there is a real question of how many of the groups of organisms we find in the world are superorganisms, and thus groups, if this concept is to specify what a group is (and thus as what group selection operates on). As we saw in Part Two in discussing putative corporate organisms, the paradigm putative example of a superorganism is an insect colony, and while Wilson has at times advocated that human social groups are superorganisms, even this extension of the application of the concept is hedged. It is, moreover, difficult to see how many of the groups on which group selection acts could be superorganisms, particularly if superorganisms are a *product* of group selection, rather than the raw materials on which group selection operates.[29]

In short, superorganisms are, at best, special types of groups, not groups per se. In Lloyd's terms, they are not only interactors, but also the beneficiaries of selection and manifestors of adaptation. Wilson and Sober's puzzling, occasional identification of groups and superorganisms may be a result of their frame-shifting approach to the levels of selection. They claim that "the organ-organism-population trichotomy can be frameshifted both up and down the biological hierarchy," using this claim to identify populations as organisms, group-level organisms. In terms that are more familiar in the levels of selection debate, this is to say that one can treat any one of genes, individuals, or groups as the kind of thing its neighbor is in this biological hierarchy, and thus that groups can be treated *as* our paradigmatic biological individuals, organisms.[30]

Where such a frame-shifting gloss on the multilevel perspective on selection goes wrong is that it mistakes an isomorphism between the conditions governing two processes – group selection and individual selection – for a substantive claim about the nature of groups. The parasitic response I have advocated to the problems of arbitrariness and ephemerality, by contrast, simply exploits that isomorphism.

Paralleling this parasitic strategy of argument with respect to groups, we can understand a group's *fitness* in much the way that we understand an individual's fitness, namely, in terms of *its* reproductive success. As with individuals, in the case of groups we also have two types of reproductive success: survivability or viability, which is the ability of a group to endure over time, and fertility, the ability of the group to produce offspring. In the case of groups, these two aspects of fitness may sometimes be difficult to distinguish, since group identity is not as sharp as individual identity.

There are two forms that the production of offspring could take: either it is the production of more individual organisms who remain members of the group (growth), or it is producing more groups, containing individual organisms (dispersal). Both growth and dispersal involve producing more individual organisms, but they are in principle independent means of increasing the fitness of the group such that groups with the given phenotypic or group trait successfully compete with groups without it, and thus come to replace those groups in the metapopulation. Thus, there is (group) selection for groups with the relevant trait, in the same way that there is (individual) selection for individuals with the relevant trait within populations.

This implies that a phenotypic trait (that is, a trait of individual organisms) could increase its representation in the metapopulation via group selection in either of two ways. First, it could do so by the differential addition of individual organisms to existing groups – paradigmatically by one group increasing in size, or a competitor group having its size decreased. Second, it could increase its metapopulational representation by the differential addition of groups of individual organisms with that trait – paradigmatically through differential colonization and migration rates between groups. The same would seem to be true of group-level traits. What is crucial to group selection is that there be differential levels of fitness between groups with respect to either an individual-level or a group-level trait.

This highlights the point made earlier that, contrary to Williams's Matching Principle, group selection could itself operate on individual-level traits, something also implied by the invocation of group selection to explain the evolution of altruism, an individual-level trait. Groups might be interactors, but individuals the beneficiaries of selection and the manifestors of adaptations. This relationship between the level at which the process of natural selection operates and the level of the traits that are selected by that process complicates the inferences between claims about the level at which selective processes occur and claims about group adaptations, but it seems mandated if we are to have a conception of group selection that better captures the complexities of the dynamics of the action of natural selection.

9. SPECIES AND CLADE SELECTION

To complete this introduction to groups as agents of selection, I want to make some brief comments about two types of group that have sometimes been singled out as special by paleontologists: species and clades, monophyletic groups of organisms or species. Paleontologists have been especially interested in accounting for large-scale patterns of evolution, and Steven Stanley and Stephen Jay Gould have argued over the past twenty or so years that some of these patterns are due to species or whole clades being the agents of selection. Although species and clade selection will not be my focus in the remaining two chapters, there are features of the debate over the status of such higher-level agents of selection that can be illuminated by our discussion thus far.[31]

There are two strands to, or motivations for, species or clade selection. The first has already been alluded to, the existence of patterns of evolution best explained by viewing species as the agents of selection. The clearest cases of such patterns are those that involve what I have been calling group-only properties. Included here are properties such as population density, which cannot be determined simply by measuring organismic traits, and sex ratios, which while a group-only property, can be so determined, and so can be viewed as aggregates of organism-level traits. For species or clade selection, the relevant groups are whole species or clades. Where there are large-scale evolutionary trends that involve such properties, the issue of what is responsible for those trends arises, and since these are group-only properties, the claim is that if they are the result of natural selection, it can only be natural selection operating on the corresponding groups, species, and clades. If there is differential

TABLE 8.1. *Individual selection and species selection*

Process	Microevolution	Macroevolution
Unit of selection:	Individual	Species
Source of variability:	Mutation/recombination	Speciation
Type of selection:	Natural selection	Species selection
	A. Survival against death	A. Survival against extinction
	B. Rate of reproduction	B. Rate of speciation

Source: Modified from Table 7.2 of Steven Stanley, *Macroevolution* (San Francisco, CA: W.H. Freeman, 1979), page 189.

extinction across clades, for example, then that cannot be explained by appeal to a process that operates only within local populations, for extinction is a property of species and clades, not individual organisms. Some, following Elisabeth Vrba, have required that "for species selection the differential prospects for speciation and extinction must depend on properties of the species themselves, rather than properties of the individual organisms that compose the species."[32]

Examples that have been claimed to fit this pattern include the evolution of planktotrophic mollusks in the late Cretaceous via their having the heritable, group-only property of greater geographic range, and so greater longevity; the evolution of larger body size in mammals via the group-only properties of population density and geographic range; and the predominance of flowering plants through the property of vector-mediated pollen dispersal promoting greater rates of speciation. In each case, there are distinctive patterns within and across clades concerning speciation rates, the duration of species, and the species and clade diversity, and the claim is that they are due to agents of selection that are lineages, rather than organisms or even mere trait groups.[33]

The second strand or motivation develops the analogy between the role of organisms in microevolutionary change, and that of lineages, such as species or clades, within macroevolutionary change. For example, consider the following table (Table 8.1) redrawn from Steven Stanley's *Macroevolution: Pattern and Process.* As Stanley makes clear in the text, the analogy here holds between species and asexually reproducing individuals, and the analogy receives an extended treatment by Stephen Jay Gould, summarized by Gould in a three-page table as "the grand analogy." By assimilating species or clades to organisms, proponents of species selection ensure that their form of group selection avoids both the ephemerality and arbitrariness problems.[34]

In terms that I have been using throughout this chapter, Steve Stanley has emphasized that species and clades need neither be the beneficiaries of selection nor the manifestors of adaptation. Thus, it is not surprising that many of the examples that he provides (for example, predation patterns within marine invertebrates) do not involve group-only but multilevel properties. Even if cases involving group-only properties represent a clear case of species selection – a point to which I shall return in a moment – as the philosopher Todd Grantham has argued, it seems a mistake to restrict species selection to such cases. If we take the organism-species analogy seriously in light of our earlier discussion, we can see why.[35]

Organism-level selection may influence the selection of organismic properties, but given the rejection of the Matching Principle, it need not. Just as organisms being the agents of selection can produce certain properties in groups of organisms, a process in which species are such agents may result only or primarily in the redistribution of organismic properties. For example, in the much-discussed example of planktotrophic mollusks, suppose that heritable variation in geographic range is the group-only property on which natural selection acts. Thus, species or clades with this property are the agents of selection. Yet at least the chief beneficiaries of selection and manifestors of adaptation through this process could be individual organisms, with those feeding on plankton and so traversing greater distances being differentially represented in later generations, and these individual-level properties becoming adaptations as a result of species selection.

This relates to the distinction that Elisabeth Vrba has drawn between species *sorting* and species *selection*. "Species sorting" refers primarily to the pattern of traits distributed across species, this being the outcome of natural selection or other forces. There is a (derivative) process of species sorting whenever such a pattern is identified. Species selection, by contrast, is one possible process generating that pattern. As Kim Sterelny has noted in a recent discussion, this distinction is recognized by all parties to the discussion over species selection, and it applies to both species and clades. One of the central issues in the debate over species and clade selection is how to distinguish between cases in which we have mere species sorting, and those in which the sorting for which we have good evidence is likely to be caused by a process of species selection, rather than simply an effect of either selection operating at other levels or of nonselective processes (for example, genetic drift). Vrba's own view is that many putative cases of species selection only exemplify species sorting, in large part because of her insistence that species selection requires (in our terms) group-only properties of species.[36]

The distinction between sorting and selection is applicable to any putative case of selection, not just that of species and clades. In effect, it is equivalent to Elliott Sober's distinction between the selection *for* properties and the selection *of* either correlative properties or objects that happen to have the properties selected for. (This is so in spite of Vrba's own antipathy to Sober's use of that distinction.) If species are merely sorted, if there is selection of them but not for the properties they themselves have, then they may be the beneficiaries of selection but they are not agents of selection.[37]

Given the collapse of the Matching Principle, and the concerns that I have raised about the No Higher Than Principle, I am skeptical of the work that the distinction between sorting and selection can do in the debate over higher-level selection. It is not simply that our empirical data, however ingeniously collected, may be insufficient to allow one to distinguish the two in particular cases – although that is surely often the case. Rather, I think that the distinction is harder to draw *conceptually* once we turn to the details of those cases.

Suppose that we grant the strictest criterion for species or clade selection: that there be species- or clade-only traits, and that these be nonaggregative traits, like population density or geographic range. In terms of the distinction between sorting and selection, there must be a pattern of sorting that is caused by, or depends on, the selection of such properties, *rather than* by or on properties of lower-level agents, such as organisms. But in a biological world in which the dependencies between the various hierarchically ordered biological agents are multidirectional, whenever there is such high-level selection, there will almost certainly also be some sort of lower-level selection. This is not least because even nonaggregative, group only properties do not float free of the properties of organisms in those groups. The mechanism of natural selection has a limited fineness of grain, and where properties at different levels are not merely correlated but causally integrated, it makes little sense to talk of one *rather than* the other being selected, or one rather than the other being sorted.

What I am suggesting is that a distinction that is "conceptually clear" in simple and paradigm cases loses that status when we try to apply it to an intricately ordered, messy biological world in which properties both at a level and those between levels face the tribunal of selection together. The distinction between sorting and selection, like that of the selection "of" and "for" certain properties, presupposes a separation between the levels of selection that is considerably neater than we are likely to find in any real-world case.

The opening moves made in this chapter will allow us to take a deeper look at group selection in the remainder of Part Four. In the next chapter, I continue with an extended examination of a putative, often-discussed case of group selection "in the wild," that of the evolution of the myxoma virus in populations of rabbits in Australia from the 1950s on. The final sections of that chapter discuss some broader issues that this case study raises – about pluralistic views of natural selection, the idea of levels of selection, and the seemingly ineliminable role of perspectives in characterizing natural selection – that we will explore further in Chapter 10.

9

Arguing about Group Selection

The Myxoma Case

1. GROUP SELECTION IN THE WILD

In the previous chapter, I mentioned the distinction between earlier forms of group selection, such as those associated with the Chicago School of ecology and V. C. Wynne-Edwards, and more recent forms of group selection stemming from David Sloan Wilson's work on trait groups. One of the contrasts between these two traditions lies in the degree of mathematical rigor within each. The mathematical modeling of group selection that currently exists itself has followed two relatively distinct traditions. The first is a laboratory or experimental tradition with its roots in Sewall Wright's work on evolution in structured populations and exemplified by Michael Wade's work over the past 25 years at Chicago and Indiana. The second is an adaptationist tradition that, while keeping "old" approaches to group selection at arm's length, has sought to use its mathematical sophistication and philosophical savvy to overturn the influential challenges to group selection issued by the rise of the gene's eye view of natural selection.[1]

One limitation that both traditions face is that they can be seen as "merely theoretical" by their opponents. Indeed, despite attention within these traditions to the practical or applied side of their theoretical work, such a perception is common amongst evolutionary biologists who are not specialists in the area. Group selection might be brought about through artificial means in a laboratory, or might receive an adequate, robust mathematical and philosophical justification, but what about the real world? What is the evidence that groups are the agents of selection in the wild?

A standard example often given of group selection in the wild is the spread of the myxoma virus after it was introduced in Australia in 1950 in order to control the rabbit population there. The virus causes myxomatosis, a disease that can prove extremely lethal, killing rabbits often within a matter of days. Richard Lewontin introduced the myxoma case as an example of group selection that was likely to be widely instanced, and this view of the evolution of myxoma has been developed further by Elliott Sober and David Sloan Wilson. Those skeptical of group selection have provided alternative interpretations of the case. Sober and Wilson have replied to some of these, yet despite my sympathies with many of the views that Sober and Wilson defend, I also think that what one should say about the myxoma case is less clear cut than Sober and Wilson seem to think it is.[2]

The example can also be used to raise some general issues about the nature of selection and how it is usually conceived. One such issue is the place of test cases in science, and the resolvability of scientific disputes, such as that over the agents of selection.

2. TEST CASES AND RESOLVABILITY

By virtue of their controversial or novel nature, some views and theories carry a special burden of proof. A dialectically powerful way to discharge that burden is to present a clear and decisive example that can only or can best be explained by the view or theory. In science, such examples take the form of phenomena that serve as test cases for the view or theory under consideration.

The theory of group selection is one such view. As I mentioned in the previous chapter, David Sloan Wilson and Elliott Sober have been at the forefront of the revival of group selection within evolutionary biology. Amongst the theoretical work that they have done in articulating group selection as a coherent and potentially powerful force directing evolution, and the empirical evidence they have adduced in support of the efficacy of group selection in actual populations, are several examples they represent as test cases of their theory. In their recent *Unto Others*, they present two such examples: the evolution of female-biased sex ratios, and the evolution of virulence.[3]

In this chapter, I shall focus exclusively on the latter case, particularly on the most detailed example they discuss, that of the evolution of avirulence in the myxoma virus. There are two immediate conclusions that I shall argue for from an examination of this case, and a third that warrants further discussion. These conclusions carry with them a number of broader implications for how we do and might think about biological reality that I discuss further in Chapter 10. The immediate conclusions themselves are of increasing generality, moving from a weaker to a stronger claim about the myxoma case, and then in turn to the debate over the levels of selection more generally.

First, as described in the philosophical literature on the agents of selection, the myxoma case does not, contra Wilson and Sober, "provide compelling evidence for group selection." It is not the decisive test case of group selection theory that they claim it is. This is not simply because there are possible, competing individual-level explanations of the case, but, as I will argue in sections 3 and 4, because there is no independent basis for preferring the group selection account of the phenomenon to alternative explanations.[4]

Second, the myxoma case is likely to remain subject to multiple, alternative explanations, only some of which appeal to group selection. This second conclusion goes beyond the first not only in strength, but also by taking us beyond the abbreviated characterizations of the myxoma case that have featured in the debate over the agents of selection into the epidemiological and virological literatures in which myxomatosis is discussed more fully. Given standard views of how scientific disputes are rationally resolved, one could reasonably expect the addition of further empirical details to the case to reveal the true level or levels at which the evolution of virulence is governed by natural selection. Yet far from achieving this goal, restoring some of the complexities to the myxoma case usually omitted or glossed over by philosophers and biologists alike reinforces and strengthens the first conclusion. The myxoma case is likely to be irresolvable vis-à-vis the debate over the agents of selection. Defending this more general conclusion will be my focus in sections 5–7.

Third, what is true of the myxoma case is true more generally, and so the overall debate between proponents of group selection and other agents of selection is also likely to be irresolvable. Invoking the metaphor of viewpoints or standpoints, many evolutionary phenomena that can be adequately explained by invoking one viewpoint – whether it be that of the gene, the individual, or the group – can be explained adequately by invoking at least one other viewpoint. A defense of this final claim would involve at least a detailed examination of the other test case for group selection theory that Wilson and Sober provide, that of the evolution of female-biased sex ratios. And a version of this generalization that treated various levels of selection symmetrically would require exploring other phenomena, such as meiotic drive, that have been claimed by many to constitute decisive cases vis-à-vis the theory of *genic* selection.[5]

A strong form of this third conclusion is entailed by a widespread form of pluralism about the levels of selection that maintains that these viewpoints are, in an important sense, equivalent. But my argument in this chapter does not rest on pluralism about the levels of selection. Indeed, in Chapter 10, I shall explore pluralism more fully and argue against several forms of pluralism that have recently gained some currency in the literature on the levels of selection.[6]

I begin in the next section with a familiar sketch of the myxoma case and the debate over it, going on in section 4 to argue against the most commonly cited reasons for resolving this debate one way or the other. Included here is the appeal that Sober and Wilson make to "the averaging fallacy" in criticizing individualistic explanations of the myxoma case. This fallacy has been invoked by these authors, separately and together, for many years in their defense of group selection. In keeping with the general themes of this chapter, I shall argue that attributions of the fallacy do little to support the idea that the myxoma case is a paradigm case of group selection in the wild.

3. THE CASE OF THE MYXOMA VIRUS: A SKETCH AND TWO INTERPRETATIONS

The myxoma virus was introduced into Australia as a way to control an exploding rabbit population in 1950, a solution that met with unexpected limits to its success. After myxomatosis initially killed 99% of the hosts infected, and seemingly quickly, the rabbit population began to rise steadily again. As expected from individual selection theory, rabbits from the wild tested against laboratory strains of the virus had an increased resistance to the virus. But it was also found that the virus in the wild had decreased its virulence, compared to those very laboratory strains, a finding anomalous within the theory of individual selection.

In Australia in 1950, the virus was primarily transmitted between host rabbits by an arthropod vector, the mosquito. As Lewontin notes, "mosquitoes do not bite dead rabbits." Thus, the transmission of the virus from dead hosts is minimal. More virulent forms of the virus kill the host organism more quickly, and thus are less likely to be transmitted to other host organisms. Less virulent forms of the virus tend to fixate on the overall host population of rabbits, with more virulent forms disappearing. The chief phenomenon to be explained is the apparent evolution of what is often termed "avirulence" in the myxoma virus. Why did avirulent strains of the virus predominate? What was the mechanism that brought this observed result about?[7]

The standard group selectionist interpretation and explanation of this phenomenon is as follows. Within single host organisms, more virulent forms of the virus are favored, since these have a higher reproductive rate, and come to replace less virulent strains within any given rabbit. Thus, higher virulence is fitter within an organismic host than is lower virulence and evolves by individual selection. But forms of the virus with lower virulence are likely to infect more hosts by allowing their hosts to live longer, and so are fitter across organismic hosts. In effect, viruses on a given rabbit are a trait group in both senses discussed in the previous chapter: they are "populations enclosed in areas smaller than the boundary of the deme," and sets "of individuals that influence each other's fitness with respect to a certain trait but not the fitness of those outside the group." If we assume that an individual rabbit delineates a group of viruses, then this implies that low virulence is altruistic, and that it evolves because of the higher fitness of less virulent groups of viruses. This is a process of selection between groups, and thus an example of group selection.[8]

There is also an obvious, contrasting individualist explanation for the phenomenon, one provided by Douglas Futuyma and Richard Alexander and Gerald Borgia. A viral strain that kills its host is less fit than one that does not kill its host. Low virulence in a viral strain reduces the chance that it will kill its host. Thus, low virulence is an adaptation of individual viral units, and it evolves by a process of selection that acts on individuals. As a result, lower levels of virulence spread through the population of viruses as a whole via individual selection.[9]

On this view, the only relevant groups are the overall population of rabbits and the overall population of viruses. Individual selection changes the composition of the latter of these from one with predominantly highly virulent individuals to one with predominantly less virulent individuals over evolutionary time. In positing low virulence as an adaptation of individual viruses, this view also denies the existence of low virulence as an altruistic trait, since low virulence is depicted as being to the evolutionary advantage of the individual virus.

Sober and Wilson have made two related criticisms of this sort of individualistic interpretation, criticisms that have a longer history. First, it involves what they call the *averaging approach* to calculating fitness: it calculates the fitness of traits across the entire metapopulation, rather than within the constituent populations, that is, groups. Sober and Wilson regard the averaging approach to fitness in this context as involving a fallacy, the *averaging fallacy*. As a fallacy, this is the mistake of using the averaging approach to fitness to draw a conclusion about the level of selection that governs its distribution. This inference is a fallacy, because in adopting the averaging approach, one collapses the distinction between within- and between-group selection, itself crucial to characterizing, respectively, the process of individual selection and that of group selection. Thus, once one adopts the averaging approach, no inferences can be validly drawn about the level at which selection operates. Second (and subsequently), by focusing exclusively on the products of natural selection – overall fitness values – this sort of interpretation ignores the processes that generate those products. The debate over the agents of selection, however, is precisely one over how natural selection operates. Thus, the individualistic explanation begs the question at issue.[10]

Individualistic explanations of the myxoma case have been motivated by a general concern over the coherence or prevalence of group selection, one that turns, as I indicated in the previous chapter, in part on doubts about the ontological status of groups. Recall the arbitrariness problem: that there is no way to demarcate groups of organisms from mere sets of organisms, thus implying that any n-tuple of individuals could be said to constitute a group. And the ephemerality problem: that groups, unlike individuals or genes, are not sufficiently permanent arrangements in the biological world to be the units on which natural selection operates over evolutionary time. Both problems reflect a concern over the ontological status of groups insofar as they express the view that groups are "not real" in the way that genes and individuals are. Such existence as they have makes them unsuitable to play the role of an agent of selection in general, and in the myxoma case in particular.

4. THE STATE OF THINGS: AVERAGING AND THE STATUS OF GROUPS

If the individual selectionist account of the myxoma case were to commit a fallacy, as Sober and Wilson claim, then clearly that would provide a reason to favor the group selectionist account of it. And, conversely, if the group selectionist account of the myxoma case were to suffer from either the arbitrariness or the ephemerality problem, then the apparent gridlock between the two explanations would be broken in favor of the individual selection account. It is important to recognize that neither resolution of the debate over the myxoma case is plausible.

Consider first the claim that the group selectionist account of the myxoma case suffers from the arbitrariness and ephemerality problems, and that this provides a conclusive reason for preferring the individual selectionist account. Neither charge holds of the groups of viruses bounded by individual host organisms. A rabbit-bound population of viruses constitutes an entity that faces shared ecological and evolutionary circumstances. Individual viruses in such a group draw on the same resources, and by virtue of physical proximity to one another, face the possibility of host resistance or vector transmission together. Importantly, populations of viruses on different rabbits are likely to differ from one another in these respects, and so differ in fitness. Thus, rabbit-bound groups of viruses are not arbitrary from an evolutionary point of view (cf. section 6 that follows, however). Furthermore, given that typically there will be hundreds of generations of viruses on any given rabbit, these groups are not ephemeral either. In short, even if there *are* putative instances of group selection that are faced with the arbitrariness and ephemerality problems, group selection in the myxoma case cannot be rejected because the case invokes arbitrary or ephemeral groups.

Consider now the averaging fallacy. There are two responses that can be made to the charge that such a fallacy is committed in the individual-level explanation of the myxoma case. The first of these concedes that the inference from adopting the averaging approach to fitness to a conclusion about the level at which selection operates is a fallacy, but denies that this inference is made in explaining the myxoma case. The second response argues that the attribution of the averaging fallacy in this case, and perhaps more generally, itself begs the question against the proponent of individual selection. Let us take each in turn.

If the averaging fallacy is committed in the individual level explanation of the myxoma case, then that explanation must somewhere invoke the averaging approach to calculating fitness. Thus, whether the averaging fallacy is committed turns on what basis there is for the claims that:

1. A viral strain that kills its host is less fit than one that does not kill its host.
2. Low virulence reduces the chances of killing one's host.

Thus,

3. Low virulence is an adaptation of individual viruses.

So,

4. Low virulence evolves by individual selection.

For the averaging fallacy to be committed here, the fitness explicit in (1) and implicit in (3) must be conceptualized or calculated as an average over the whole population of viruses. Furthermore, this averaging must be the reason for holding at least one of (1)–(4) to be true.

There are, however, at least two related reasons for thinking that (1)–(4) are true that are independent of whether fitness is calculated over the whole population. Both reasons pack some dialectical punch in part as a result of the distance that proponents of the "new" group selection have self-consciously promoted between their views and the "old" group selection tradition featuring Emerson and Wynne-Edwards.

First, it has been well known, at least since the work of David Lack in the 1950s on optimal clutch size in birds, that individuals can increase their reproductive success by limiting the number of offspring they produce. One might think that reduced virulence in individual viruses is simply an instance of such a reproductive strategy, and so an adaptation of individual viruses that reflects the increased fitness of those viruses. Explanations such as Lack's have not been seriously disputed by proponents of group selection as appropriate in at least some cases. This suggests that it does not commit the averaging fallacy. Given that, there would have to be some relevant difference in the myxoma case that vitiates the same reasoning that Lack used, were any "averaging fallacy" charge to be plausible in that case.[11]

Second, since the fitness of any trait needs to be relativized to an environment, low virulence may be viewed as having higher fitness once we consider an environment in which host death constitutes a major limitation on long-term reproductive success. One comes to view lower virulence as an adaptation by considering the nature of this environment more fully, rather than simply by defining fitness and adaptation as "what evolves." Again, there is a basis for (1)–(4) that appeals to points that group selectionists themselves appear to be committed to.

Such reasons are independent of whether we calculate fitnesses within or across local groups of viruses (= rabbits) in that they are factors "in play," so to speak, prior to any calculations of fitness at all. Of course, if the individual selectionist viewpoint is adopted, then fitnesses *will* be calculated across the whole population of viruses. But provided that this is not the basis for adopting (1)–(4), the averaging fallacy has not been committed.

So much for the first response to the averaging fallacy charge. The second response is more dialectically powerful in the context of the overall argument of this chapter. The idea behind the averaging fallacy charge is that individualists illegitimately collapse the distinction between within- and between-group fitnesses by averaging across groups to begin with. But one can collapse this distinction (illegitimately or not) only if the distinction is already there to collapse. This in turn presupposes not simply that there are groups of viruses (= rabbits) in the overall population, but that these groups constitute a causal structure to which natural selection itself is sensitive. Yet this is precisely what proponents of the individual level explanation will deny – not necessarily in general but in this particular case. For although there is a sense in which rabbits are groups of viruses, it is not this group structure on which natural selection operates. Rather, it operates on individual viruses, viruses whose fitness is sensitive to features of their environment, including the nature of their hosts (= groups of viruses) and limitations on transmission imposed by the nature of the vectors that mediate the spread of the virus. In short, an averaging fallacy can be committed only if we already concede that groups themselves are acted on by natural selection, and this is precisely what is at issue between proponents of the two explanations. Ironically, it is Sober and Wilson's attribution of an averaging fallacy itself that begs the question against the individualist.

To summarize: Sober and Wilson argue that at least some individualistic views of the myxoma case commit the averaging fallacy, and that interpretations that rely crucially on a misplaced skepticism about the reality of groups (of viruses, in this case) should be rejected. I concur. But in generalizing this claim and presenting the myxoma case as a test case for the theory of group selection, a test that the theory passes, Sober and Wilson are mistaken. The test can also be passed by the theory of individual selection. Moreover, their general invocation of the averaging fallacy itself begs the very question at issue. At this point, I draw my first conclusion: that the myxoma case, as it is usually presented, is not a decisive example of group selection.

I have already remarked that the dispute over the myxoma case has involved an abbreviated characterization of the case. One healthy suspicion fostered by interplay between the philosophy, history, and social studies of science over the last forty years is directed at just this sort of impoverished treatment of actual examples. Might our anemic characterization of "the evolution of avirulence" in myxoma actually create or maintain the standoff in the debate over it? Shouldn't a more complete rendering of the facts allow a resolution of the apparent deadlock between proponents of our two perspectives?

In the next section, I fill out the picture of the pathology of the disease and the nature of its transmission, drawing largely on the masterful book-length treatments of Fenner and Ratcliffe and Fenner and Fantini. I shall argue that not only does a more complete rendering of the case fail to challenge my conclusion about the irresolvability of the case, but that it actually reinforces that conclusion in a way that suggests the pair of stronger conclusions about irresolvability stated in section 2.[12]

5. A LITTLE MORE ON THE VIROLOGY AND EPIDEMIOLOGY OF MYXOMATOSIS

The myxoma virus is a member of the family *Poxviridae*. The disease it causes in rabbits was first described by the Italian medical researcher Guiseppe Sanarelli in 1898 when European rabbits he received for his laboratory work in Uruguay (sent from Brazil) developed the disease. The virus is relatively host specific, being largely restricted to a variety of "rabbit species," including those from the *Sylvilagus* and *Oryctolagus* genera, commonly called, respectively, the Californian and European rabbit. The strain of the virus introduced into Australia was isolated from an *Oryctolagus* rabbit in Brazil in 1911. This strain was maintained through serial passage from host to host until it was introduced to Australia in 1950.

At the initial site of myxoma infection in rabbits a skin lesion appears, and the virus begins replicating in the body of the host, working its way through lymph nodes and the circulatory system. Secondary lesion sites also appear, and there is often a discharge from the eyes and a swelling of the genitalia. The preferred concept of virulence used by epidemiologists is host-fatality or lethality, the percentage of infected rabbits that die, though this is commonly estimated from a more readily operationalized measure, the average number of days to death. An estimate was necessitated in Australia by the impracticality of keeping large numbers of rabbits, and was made possible by the apparent correlation and regression coefficient between case-mortality rate and the more readily measured-in-the-lab variable of average number of days to death.

With the original strain of the virus introduced into Australia, infected laboratory rabbits survived about five days, making the virus extremely lethal. In one of the first systematic field counts, at Lake Urana in 1951, the lethality of the virus was estimated at 99.8%, this being reduced to 90% in 1952.[13]

TABLE 9.1. *The virulence of myxoma strains sampled in Australia, 1950s*

Virulence grade	I	II	III	IV	V	
Degree of virulence	Extreme	Very high	Moderate	Low	Very low	
Mean survival time (days)	<13	13–16	17–28	29–50	–	Number of strains tested
Case-fatality rate (%)	99.5	95–99	70–95	50–70	<50	
1950–51	>99					
1951–52	33	50	17	0	0	6
1952–53	4	13	74	9	0	23
1953–54	16	25	50	9	0	12
1954–55	16	16	42	26	0	19
1955–56	0	3	55	25	17	155
1956–57	0	6	55	24	15	165
1957–58	3	7	54	22	14	112
1958–59	0.5	20	57	14	8	179

Source: Modified from Table 7.2 of Frank Fenner and Bernardino Fantini, *Biological Control of Vertebrate Pests: The History of Myxomatosis – An Experiment in Evolution* (Oxford: CABI Publishing, 1999).

Because the concept of virulence used here is relational, this reduction in virulence alone could in principle be accounted for by a change in host resistance in the population. But suspecting that the virus had changed, researchers checked virus serum taken from infected rabbits against captive rabbits that had been isolated from the virus until that point. They found, indeed, that the virus itself was reduced in its lethality. In their attempts to measure the changes in the virulence of the virus itself, from 1951 researchers began gathering multiple samples of the virus from wild populations of infected rabbits. By 1959 they had 672 such strains (no doubt, not all distinct), and they introduced five grades of virulence for these, rated by their estimated case-mortality rates, starting with >99% for Grade 1, the original strain (rarely recovered from the wild), through to <50% for Grade V. The virus stabilizes at Grade III, with this constituting roughly 50% of the strains throughout the 1950s, with an estimated lethality of 70–95%. As can be seen in Table 9.1, the incidence of Grade II viruses (lethality: 95–99%) remains over 10% of the sampled population for much of the period surveyed, and increases significantly from 1958 to 1959. The incidence of Grade V viruses remains extremely low throughout, and Fenner and Ratcliffe note that "it is unlikely that a virus of this type would survive in nature." Simple talk of "the evolution of avirulence" neither conveys nor does justice to the complexities here.[14]

There is a further complication to the basic finding that less virulent strains have evolved. The system of viral grades, the categorization of particular strains under them, and the resulting distribution pattern that changes over the years, all presuppose the correlation between mean survival time and survival rate mentioned a few paragraphs ago. Work led by Ian Parer has recently challenged this presupposition, and called into question whether the change in virulence was as radical as suggested by data such as that represented in Table 9.1.

First, Parer and his coworkers found a sire effect of 20–25% in off-spring immunity, where sires had been infected with myxomatosis less than ten months prior to the birth of their offspring. The myxoma virus is disproportionately accumulated in the testes of wild rabbits, with its DNA remaining there long after infectious myxoma virus is no longer found there (or in other tissues). As they say, "the sire effect is likely to have contributed to the observed initial rapid increase in the resistance of rabbits," and to account for some of the variation in measured virulence from year to year.[15]

Second, Parer points out that the use of mean survival times as a measure of virulence has systematically underestimated the lethality of the myxoma virus. Parer was prompted to investigate the relationship between mean survival time and survival rate (lethality) after finding that most field strains of the virus were of high lethality (Grade I). Basically, the initial regression coefficient between the two was artificially high, due to the inclusion both of one highly attenuated strain of virus and of estimated survival times for rabbits surviving beyond fifty days. Once these are excluded and the relationship between mean survival time and lethality recalculated, the regression slope for these two variables shifts from .67 to .23. This means that although the correlation between the two remains strong, small variations in mean survival time correlate with large variations in lethality, and so the small errors in the former can lead to large errors in the latter. Parer argues that relying on an inflated regression between mean survival time and lethality "has resulted in most field strains being allocated inappropriate grades and it has distorted to some extent our perceptions of the types of evolutionary changes that occurred after the myxoma virus was introduced into Australia." The full significance of these findings remains subject to further investigation.[16]

Transmission is also more complicated than the standard picture from section 3 suggests. While the most prevalent and far-reaching form of transmission is via an arthropod vector, the virus can also be transmitted through direct contact or contagion between hosts, or through nonanimate vectors, such as thornbushes or warrens. Given that rabbits live in close proximity to one another in hutches, these are actual (not merely possible) mechanisms for transmission. Arthropod vectors pick up the virus on their mouthparts when they feed on an infected host animal, particularly at a lesion site, and then transmit the virus when they move to another host, or to another part of the same host. Vectors preferentially feed on the head of the rabbit, perhaps because it is the primary site of lesions. Thus, as well as between-host transmission of the virus, there is within-host transmission facilitated by vectors themselves and by the host's circulatory system, and differential vector transmission within a host.

Transmissibility of the virus by arthropod vectors, such as the mosquito, depends on facts about the ecology and life cycle of those vectors. The concentration of mosquitoes varies in accord with heat and moisture, and thus their abundance follows a seasonal cycle. This represented a problem for the use of myxoma in controlling the wild population of rabbits in Australia, not only in areas in which the concentration of mosquitoes was insufficient to transmit the disease effectively at any time of the year (for example, in the drier plains of Western Australia), but in all areas during the winter months. In the late 1960s, in response to this problem, the Australian government explored possibilities for introducing another vector for transmission, the European rabbit flea (*Spilopsyllus cuniculi*) that infests rabbits and does so year round. *S. cuniculi* was, in general, an effective vector for transmitting the myxoma virus throughout the year, and was particularly effective in infecting young rabbits, which significantly reduced the population growth for the following year. However, *S. cuniculi* survived poorly in arid conditions, and in the early 1990s another flea species, *Xenopsylla cunicularis*, was introduced from Spain.

The introduction of vectors alternative to the naturally occurring mosquito was thus aimed primarily at solving what has been called the *problem of overwintering*. Due to vector reduction during the winter months, more virulent forms of the myxoma virus could not survive the winter: they killed their hosts before being spread to other hosts. Fenner and Ratcliffe identify the occurrence of attenuated strains in geographically distinct areas as a reason to think that the virus was mutating within local populations, rather than being vector-transmitted there, since it was unlikely that vectors could reach these areas from regions in which attenuation had already taken place. The switch in vectors met, however, with mixed and limited success as a way to control the rabbit population, a point I shall return to in the next section.[17]

Finally, given that the attention commanded by the myxoma case amongst those interested in the mechanisms of *natural* selection derives in part from its "in the wild" status, it is relevant that the bulk of the long-term data reported is the result of the natural overwintering of the virus in rabbit populations together with supplemental "inoculation campaigns." This program of human-induced infection was systematic and sustained, involving direct transmission of the virus by human beings (researchers and farmers) to over 20,000 rabbits on average in each of the twelve years following its original introduction. In addition, farmers intervened in another way in engaging their fondness for shooting and poisoning rabbits, which had become a major threat to their livelihood. Thus, human agents play a key role in the spread and maintenance of myxomatosis, not just through the selection of additional vectors, but also by serving as initiating vectors for transmission of the virus and in culling the (infected) rabbit population. This is what one would expect in a case that is, after all, a form of pest control. Whether these facts impugn the status of the myxoma case as one of group selection *in the wild*, I leave as an exercise for the reader.[18]

6. WHY FURTHER DETAILS REINFORCE THE IRRESOLVABILITY CLAIM

There is a surprisingly large gap between the epidemiological literature on myxomatosis and the evolutionary debate that the phenomena it studies has inspired. The tendency in virology and epidemiology is to focus on the documented findings in natural populations of rabbits via sampling techniques, approaching the question of the mechanisms mediating changes here through laboratory studies of both rabbit and virus. There has been little explicit attention to the question of just how natural selection operates in this case. For example, Fenner and Ratcliffe attribute the reduction in virulence they document as "being due to mutations of the virus which were subsequently selected for," but do not specify the selective mechanism. Later, in discussing the evolution of attenuated strains of the virus, they appeal to both the greater survival over the winter months of rabbits with attenuated strains of the virus and to lower temperatures during the winter as facilitating this survival (of both rabbits and virus). Again, the nature of the selective mechanism itself is not further specified. In the extensive overview that Fenner and Fantini provide of the history of myxomatosis, this question of the level at which selection operates remains not simply unanswered, but (to a close approximation) unasked.[19]

General textbook treatments of the example have sometimes taken the "classic view" of the evolution of lower levels of virulence to be group selectionist. I have found it difficult, however, to find such a view – indeed, any explicit view on the agents of selection issue – in Fenner and Ratcliffe and Fenner and Fantini. Those involved most directly in the field and laboratory work on the myxoma virus seem to have individual selection, if anything, in mind as the form that natural selection takes. This is largely because of the emphasis on a range of factors that affect the spread of lower levels of virulence – host resistance, temperature, means of transmission – all discussed in terms of their effects on individual fitness.[20]

Consider, by contrast, the literature on the levels of selection that appeals to the myxoma example. Typically describing the example in a short paragraph or so, the details sketched previously are omitted. In particular, transmission by contagion is ignored, as are secondary lesion sites and preferential vector feeding, the effect of which is to highlight the significance of vector-mediated rabbit-to-rabbit transmission, that is, transmission between groups of viruses. For group selectionists, population structure "pops out" as a salient cause of the spread of what is observed to evolve, that is, lower levels of virulence. By contrast, although individualists ascribe a role to population structure, as we will see later in this section, they oppose this interpretation on quite general grounds. They subsequently see little need to appeal more than minimally to the details of the case, being concerned primarily to debunk the claims of group selectionists.

A range of putative examples of decisive empirical findings, actual or possible, is suggested by the details sketched in the previous section. These include those from studies that artificially increase (or decrease) the respective strengths of individual and group selection in an experimental group, and the examination of evolutionary patterns of viral replication both within groups (rabbits) and between groups. The basic problem with such scenarios being viewed as potentially decisive resolutions to the debate, however, is that it is fairly easy to construct explanations from either of these perspectives for whatever data is found or even merely envisaged. Lest this sound a little too *a priori* as a pronouncement of what we must find, let us consider a few good candidates for just the sort of decisive result we seek. Since the primary interest in the case is whether it constitutes a decisive test for group selection, I focus on results that might be taken to offer such decisive support.

For example, consider the basic finding that virulence increases within hosts but decreases over the whole population over time. Surely, one might think, this at least prima facie supports the group selectionist view, since that view posits opposed forces of selection at different levels that directly explain the within- and between-group trends. Yet an individualist should respond that while individual selection should indeed increase virulence, faced with the environmental circumstances of shorter-lived hosts, it will also drive lower levels of virulence. That is, rather than seeing individual and group selection as competing forces that pull in different directions, as does the proponent of group selection, the individualist views both forces as being at the individual level. Each reflects a different adaptive pressure faced by individuals, with the one driving lower virulence winning out.

Likewise, consider experimental or naturalistic conditions that putatively reduce the effects of group selection and that thus lead to increases in levels of virulence. Lewontin had predicted that the introduction of the European rabbit flea, *S. cuniculi*, would increase virulence, presumably because doing so would effectively reduce the "groupishness" of viruses located on particular rabbits, and so limit the effects of group selection in promoting reduced levels of virulence by increasing its transmissability between rabbits. Prima facie, such manipulations would support the group selectionist explanation, in much the way that the findings of Michael Wade's celebrated flour beetle experiments that used an artificial group selection paradigm support group selection. In both cases, we manipulate factors that would adjust the strength of the force of group selection, were it to exist, and observe a phenotypic change that confirms the hypothesis that group selection is acting in the natural environment.[21]

There are two problems with this idea, one pertaining to the inconclusiveness of the empirical data relevant to Lewontin's particular suggestion, the other more general.

The first is that the effects on the virulence of the myxoma virus of introducing an alternate vector remain unclear. In part this is due to regional variation and the environmental sensitivities of this new vector. Studies at government sites around Canberra in 1977 and at Lake Urana in New South Wales in 1981 suggested that the introduction of rabbit fleas had little effect on the transmission of the myxoma virus, while results from South Australia in 1983 supported the effectiveness of the new vector in transmitting the virus. Temperature and moisture are crucial variables here, and the mxyoma virus continued to be transmitted by the preexisting vector, the mosquito, in all three locations over the summer months.[22]

The second and deeper problem is that even had the empirical results been more clear-cut, they could also have been explained from the individualist's point of view. This is because the relevant manipulation changes the nature of the population-structured environment in which individual organisms exist, and with it the relationship between individuals and this environment. The explanation encapsulated in (1)–(3) would no longer apply, true enough. But that is because in the new environmental circumstances killing one's host is no longer detrimental to the fitness of the individual virus. So low virulence loses the fitness advantage it had in the pre-flea environment. In the new environment, increased virulence has a higher level of fitness, and so we would expect it to be the trait that evolved.

Consider a third candidate finding that might be thought to provide the decisive evidence sought. It is possible for a microorganism to have a high reproductive rate without killing its host. *Escherichia coli* bacteria, for example, are extremely rapid reproducers in the human gut, but do not usually pose a threat to the life of the host organism. In the case of the myxoma virus, there could be a strain of the virus that approximated *E. coli* in this respect, somewhat like how myxoma actually operates in *Sylvilagus* hosts. For example, a viral strain that increased the number of lesions on a rabbit, thus increasing transmissibility, with reduced infection of vital organs in the host (for example, due to host adaptation), would have this effect. Likewise, a strain that concentrated lesions on the head of the rabbit (where *S. cuniculi* concentrates, especially the ears) would have a similar transmission advantage. With an increase in transmissability, we could expect that, *ceteris paribus*, levels of virulence would increase.

I should like to make two points about the gap between this prediction and any conclusion about the level or levels at which such a selective force acts. First, the intricacy of the relationships between host, vector, and virus, and how each of these are affected by environmental parameters that are subject to change, makes it unlikely that all other things *will* be equal. Second, plausible explanations could be provided for such an outcome from each of these perspectives. Whether such an increase in virulence brought about by increased transmissability does so by virtue of selection operating directly on real population structures, groups of viruses, or does so by virtue of its operation on an individual's sensitivity to changes in its environment, would remain open.[23]

These examples indicate the sorts of problem with the appeal to decisiveness. There is, however, a principled reason why the problem here is general. We have two paradigms of evolutionary explanation, that of group selection and that of individual selection. Each has a basic repertoire of tools that can be adapted to much the same data sets in the myxoma case. Both can acknowledge the reality of a group structure (or a variety of group structures) within the overall virus population, and each accommodates this fact in its own way. The group selectionist sees natural selection as operating on the groups themselves, eliminating those groups that are less fit, which are groups whose members have high levels of virulence. The individual selectionist, by contrast, sees natural selection as operating on individuals sensitive to features of their environment, including the group structures imposed by the nature of the hosts and vectors present for transmission. If the resources of each view are rich enough to explain any putatively decisive result, as I have been suggesting, then this provides support for my second conclusion, that the myxoma case is in fact irresolvable vis-à-vis the question of how selection is really operating in it.

An aside for aficionados: it will not have escaped the notice of some that the reasoning here bears a similarity to that in Kim Sterelny and Philip Kitcher's response to the argument of Elliott Sober and Richard Lewontin that claimed that heterozygote superiority in the case of sickle cell anemia could not be explained by genic selection. Sober and Lewontin argued that, in this case, the smallest unit on which selection operates is the diploid genotype at a locus. Sterelny and Kitcher replied that individual alleles could be seen as agents of selection provided that one recognized the other allele in the genotype locus as part of the environment of the allele that is selected. If the parallel between this case and that of myxoma holds up, and what I have been arguing about irresolvability in that latter case is correct, then this debate over sickle cell anemia should also be irresolvable. End of aside.[24]

If the myxoma case is irresolvable, then it might seem a short step down the path to generalizations of this conclusion, perhaps even to the debate over group selection itself. But as I implied in the introduction, even this "short step" involves tasks that lie beyond the current chapter. For example, it would require more than the passing mention I have just made of the sickle cell case, as well as a discussion of other putative test cases of group selection, such as that of female-biased sex ratios. As a way of reinforcing my conclusion about the irresolvability of the myxoma case short of engaging in such discussions, and to move to some broader issues that this conclusion itself raises, I focus next on a metaphor occasionally made explicit in the myxoma debate and otherwise often not far beneath the surface of the debate. This is the metaphor of particular agents of selection as having *standpoints* or *viewpoints*.

7. DECISIVENESS, VIEWPOINTS, AND AGENCY

Much of the myxoma debate can be cast in terms of talk of the viewpoints of particular organisms, and indeed has been. The conception of evolution in terms of the selfish gene has perhaps made most effective use of such viewpoint talk in evolutionary biology more generally.[25]

The primary viewpoint adopted in the debate over myxomatosis is that of the virus. Group selectionists hold that the viewpoint of the group of viruses that live on a rabbit is necessary for understanding, or sheds light on, just how natural selection operates in the myxoma case. We could encapsulate the concern that individualists have about the group selectionist account in terms of whether the viewpoint of virus groups adds anything to our understanding of the phenomena to that provided by the viewpoint of individual viruses. I propose to use this metaphor in conjunction with the details sketched in the previous two sections to raise a series of further complications.

The first is that the viewpoints of viruses (individuals or groups) are not independent of those of rabbits, especially given that the central concept in play, virulence, is often characterized relationally in terms of effects of the virus on the well being of rabbits. The second is that even these four viewpoints are not exhaustive, for there is that also of the vectors, as well as that of a nested hierarchy of groups, especially of viruses. As I implied in section 5, if rabbits constitute a group of viruses, then surely (given contagious infection) a hutch of rabbits does as well, as does the local deme of rabbit hutches. And given the differential transmissability of viruses located on different parts of the rabbit, all those viruses on the rabbit's head, or its ear, or around its eyes, also constitute groups of viruses. Note that here a variation on the arbitrariness problem from section 4 arises, for although rabbit-bound viruses are not an evolutionary arbitrary group, they are far from a unique such group. Thus, the focus just on rabbit-bound viruses as the object of group selection seems either arbitrary or in need of further justification.

On the group selectionist view, within each of these groups individual selection should promote increased virulence, while group selection should promote decreased virulence. Individual selectionists, of course, have their own account of how to understand this added complexity. The third complication, in light of the first two, is that there seems to be no fact of the matter as to which viewpoint is (or viewpoints are) *the* correct viewpoint(s) for thinking about myxomatosis.

Consider now the viewpoint not of the virus but of the rabbit. When researchers found that rabbits had increased their resistance to the myxoma virus in the years following its introduction, this result was readily intelligible within the parameters of the theory of individual selection. What would have been paradoxical, in much the way as was the discovery of lower levels of virulence, would have been the finding that rabbits had decreased their levels of resistance. Although this was not found, I want to explore how this possible finding would be explained from each side of the debate over the myxoma case.

One initial thought is that if high resistance evolves by individual selection, then low resistance should evolve by group selection. (After all, this would parallel what has been said when considering the viewpoint of the virus.) To make the explanation of this explicit, consider the following argument sketch:

A. High resistance has a higher level of fitness for an individual rabbit.
B. But highly resistant rabbits spread the disease throughout the population.

Thus,

C. Groups of low resistance rabbits are fitter than groups of high resistance rabbits.

So,

D. Low resistance evolves by group selection.

Note that (A)–(D) provide an account of the evolution of low resistance in rabbits that closely parallels the group selectionist account of the evolution of low levels of virulence in the virus. In both cases, a trait that reduces within-group fitness nevertheless evolves because of group selection.

Low resistance would not, however, provide decisive support for the theory of group selection, since there is another explanation available that appeals only to individual selection. Again, to parallel the explanation of the evolution of low levels of virulence encapsulated in (1)–(3), consider the following individual-level explanation:

1.* A rabbit killed by a viral strain is less fit than one not so killed.
2.* Low resistance reduces the chances of being killed by a viral strain.

Thus,

3.* Low resistance is an adaptation of individual rabbits.

So,

4.* Low resistance evolves by individual selection.

Like the explanation of the evolution of avirulence by an appeal to individual selection, this explanation need not commit the averaging fallacy. Unlike that explanation, at least one of its premises seems strangely counterintuitive and stands in need of special justification.

This is (2*): how could low resistance increase one's fitness by reducing one's chance of parasite-induced death? Here an appeal to the sensitivity of individual rabbits to the nature of their environment needs to be made explicit. Given that the virus itself varies in its virulence, if the contraction of less virulent strains were to provide some immunization against the contraction of more virulent strains, then given other environmental conditions – such as the prevalence of less virulent strains – lower resistance could promote longevity in rabbits that have it (cf. also the sire effect, mentioned in section 5). This is how other pox viruses operate on humans, and why having chickenpox as a child is not altogether a bad thing. Fenner and Ratcliffe in fact present some evidence for just this sort of effect in myxomatosis, though they remain neutral as to whether this is caused by an interference effect between competing strains or a heightened immune response from the host.[26]

Alternatively, (2*) might be true, not by virtue of the structure of the viral environment and its interaction with individual rabbits, but because of facts about vectors and how they interact with rabbits with various levels of resistance. Infected rabbits may have less appeal for arthropod vectors than do uninfected rabbits, and so lower resistance may be a strategy that individual rabbits adopt in order to limit their exposure to highly virulent strains of the virus.

The counterfactual nature of the "finding" of the evolution of low resistance in rabbits should make it clear that this is intended as part of a "how possibly" rather than a "how actually" story. The interest of such a Just So Story is twofold in the present context.

First, it reminds us that we need to consider the full environment of an individual organism, including the nature of both its parasites and whatever group structure they have, as well as that of the vectors that mediate between host and parasite. In so doing, it draws attention to the toolkit of an individualist who appeals to the sensitivity of an individual to its (complicated) environmental circumstances. This reinforces the conclusion that such an individualist explanation will likely be available, once the empirical details of any such case are filled in, and so adds to the conclusions for which I have already argued.

Second, this "just so" story leads us fairly naturally into another – one that reminds us that there remains a further viewpoint, that of the gene. For we can now consider the following explanation of the "evolution of low resistance":

1.+ Genes whose vehicles are killed by viral strains in E are less fit than those in vehicles not so killed.

2.+ Genes for low resistance in E reduce the chances of one's vehicle being killed by viral strains.

Thus,

3.+ Low resistance is a genetic adaptation.

So,

4.+ Low resistance evolves by genic selection.

$(1^+) - (4^+)$ make explicit the thus-far implicit appeal to an environment. I suggest that the genic selectionist defending (2^+) has at least as rich a repertoire of tools as has the individualist defending (2^*).

If this is correct, then the basic problem for resolving the debate over the myxoma case is compounded, my conclusion about its likely irresolvability is further reinforced, and we have some additional reason to think that the extension of this conclusion beyond the myxoma case more generally is defensible. But this also raises two broader issues about the nature of selection and how we conceptualize it. To raise these issues as provocative questions: doesn't the irresolvability of the debate over the myxoma virus, and perhaps the levels of selection more generally, mean that the viewpoint of the individual and that of the group are simply two equivalent ways of viewing natural selection? And does the irresolvability claim here call into question the very conception of selection as operating at distinct levels? I shall say something briefly about these issues in closing this chapter.

8. HOW DEEP DOES IRRESOLVABILITY REACH?

Near the beginning of this chapter, I noted that pluralism has some vogue as a position about the levels of selection. A prominent form of pluralism, one that has proven especially popular amongst a range of biologists and philosophers of biology, holds that there is an important sense in which different models of selection are equivalent, such that the choice between them is to be made on pragmatic grounds. I shall call this view *model pluralism*, since it adopts a pluralistic stance toward various models of selection, including individual and group selection models.

Model pluralism provides a seemingly natural view to adopt if we accept either the specific irresolvability claim I have argued for about the myxoma case, or the more general irresolvability claim that I have groped toward. But despite appearances, the position I have defended in this chapter has little affinity with model pluralism. In fact, I find model pluralism prima facie implausible insofar as it implies that the disagreement between proponents of individual and group selectionist accounts of the myxoma case is merely apparent, heuristic, or pragmatic. The intuition that there is a fact of the matter about which these two sides disagree, one concerning the underlying biological ontology, runs deep. It should be given up only as a last resort.

What model pluralists are right about, and what is reinforced by our brief probe into the virological literature on myxomatosis in section 5, is that there is a deep interdependence between the various levels of selection. This presents a prima facie challenge to the predominant conception of selection as acting on distinct levels. Biological reality is perhaps more aptly conceived of as fused, enmeshed, or *entwined*, rather than hierarchically structured into neat levels, in that the properties on which selection acts, and indeed the mechanism of selection itself, do not come prepackaged at distinct levels. Conceiving of natural selection as operating at various "levels" would then be a simplification of the entwined, messy reality, a conception that imposes structure on, rather than simply reflects, biological reality. Reduced virulence in the myxoma virus can be viewed either as an adaptation of individual viruses or as a product of intergroup competition. This is not because model pluralism is true, however, but because the "levels" themselves are entwined.

The evolution of the myxoma virus in natural populations of rabbits in Australia is less conclusive as an example of group selection than proponents of group selection think. But the analysis of the example provided by their individualistic critics does not win the day either. Rather, the debate here stands as an example of a scientific dispute that cannot be resolved by the current evidence, and I have argued that it seems unlikely to yield to further evidence. Reflection on some of the complexities to the phenomenon of myxomatosis reinforce rather than ameliorate the irresolvability of the myxoma case. Thus, there are reasons to think that the scientific dispute here cannot be rationally resolved in a conclusive manner.

This conclusion almost certainly generalizes to other cases invoked in the debate over the levels of selection. While one might think that it generalizes to the whole debate between proponents of individual and group selection, that would require significantly more and I think a somewhat different argument than I have provided here. Those who endorse such a view of the general debate over the levels of selection may be tempted by model pluralism, but this is not my temptation. Rather, I think that the argument here raises questions about the adequacy of the very conception of natural selection operating at distinct levels, a conception ubiquitous in debates over how natural selection operates. If this is correct, then perhaps we need to rethink some large issues in the field.

Some of those issues concern pluralism, the notion of entwinement, and their relevance to the agents of selection. In the final chapter, I turn to these issues and the claims that I have made about them in this section.

10

Pluralism, Entwinement, and the Agents of Selection

1. THE PLURALISTIC TENOR OF THE TIMES

As I noted at the outset of Part Four, the return of the group has focused the attention of philosophers and biologists on the relationships between various putative levels at which natural selection operates, particularly on those between group, organismal, and genic selection. Both unbridled enthusiasts and more circumspect critics of the form of group selection that has received most attention – that of David Sloan Wilson and Elliott Sober – have endorsed pluralism about the agents of selection. In fact, one or another form of pluralism about the agents of selection constitutes the current orthodoxy in the philosophy of biology. The two goals of this final chapter are to critically examine this pluralistic consensus, and to introduce an alternative to it, one that questions the adequacy of the underlying conception of natural selection as operating at distinct "levels."[1]

I begin in the next section by distinguishing several forms that such pluralism has taken, turning in sections 3 and 4 to elaborate on and then critique the most widely endorsed form of pluralism and the central argument given for it. Pluralism is motivated by the intuition that there may not be a determinate answer to the question of just which level is "the" level at which selection occurs in any particular case. In section 5, I shall spell out this intuition in a way that breaks from the pluralist consensus by developing the notion of entwined levels of selection that I introduced briefly at the end of the previous chapter.

The key positive idea is that various levels of selection are often entwined or fused, not just in the sense that they co-occur, or operate in the same direction, but in that they are reliably coinstantiated and do not make isolatable, distinct contributions to the ultimate evolutionary currency, fitness. This idea calls into question the view that the levels at which selection operates are sufficiently separable for it to make sense to invoke criteria for preferring one level to others, either in particular cases or in general. I want to suggest not so much the epistemic point that it is sometimes difficult to determine whether individual or group selection is the level at which selection is operating in a particular case as to undermine the claim that there is always a determinate answer to this question. Life, unfortunately, isn't that simple.

This challenge to the idea that the biological world is neatly segmented into layers or levels has broader implications for how we might think about the metaphysics of science. Talk of the various "levels" at which objects, events, properties, states, and processes exist or can be described is pervasive in discussions within the general philosophy of science and the philosophy of particular sciences (for example, of physics, biology, psychology, economics). The metaphorical status of such talk has seldom been recognized. While there are reasons to expect entwinement to be pervasive in the biological world, and so for thinking "entwinement talk" to be more revealing than simple "levels talk" in biology, the argument of this chapter has implications for these other areas of the philosophy of science. In section 6, I briefly raise three issues for anyone wishing to explore the notion of entwinement more fully, and close by returning to two issues that have surfaced throughout *Genes and the Agents of Life*: that of smallism, and of the place of the cognitive metaphor in biological understanding.

2. TWO KINDS OF PLURALISM ABOUT THE LEVELS OF SELECTION

Consider the label "pluralism." The multilevel theory of selection proposed by Wilson and Sober has been called pluralistic, and it may pay to begin with the contrast between two forms that their own pluralism takes.[2]

The first and most obvious sense in which Wilson and Sober's view is pluralistic is that it denies that there is any single level at which selection operates; rather, selection can, and often does, operate at multiple levels, levels that often pull in opposite directions, with no one of these levels in general trumping the others. That is, there are a variety of agents of selection – gene, individual, group – with no single one being more fundamental in general than any of the others. As such, this form of pluralism suggests *realism* about the existence of the agent(s) of selection: there are, independent of our particular perspectives, distinct agents of selection in the natural world, with their own distinctive properties and subject to particular processes. And it is incompatible with what we might call *fundamentalism*, the idea that one agent of selection is more fundamental (theoretically or ontologically) than the others. I shall call this form of pluralism *agent pluralism*, since at its core is the idea that there is a plurality of agents of selection in the biological world itself.

A second and less obvious form of pluralism also exists in Wilson and Sober's multilevel selection theory. It arises in their discussion of the relationship between the theory of group selection and a variety of supposed alternatives to it. According to Sober and Wilson,

In science as in everyday life, it often helps to view complex problems from different perspectives. Inclusive fitness theory, evolutionary game theory, and selfish gene theory function this way in evolutionary biology. They are not regarded as competing theories that invoke different processes, such that one can be right and the others wrong. They are simply different ways of looking at the same world. When one theory achieves an insight by virtue of its perspective, the same insight can usually be explained in retrospect by the other theories.

Wilson and Sober go on to argue that the exclusion of the theory of group selection from this "happy pluralistic family" of alternative perspectives "reflects a massive confusion between process and perspective. The theories that were launched as alternatives to group selection are merely different ways of looking at evolution in group structured populations."[3]

This version of pluralism holds that various prima facie distinct models of natural selection – for example, selfish gene theory and group selection theory – are actually noncompeting accounts of one and the same process. At the end of the previous chapter, I called this view *model pluralism*, since it identifies a plurality in the models that evolutionary biologists adopt, rather than in the reality that those models depict. Model pluralism holds that while there may be strategic or pragmatic advantages to using one rather than another model in a particular case, these models do not compete for the truth about the nature of natural selection.

Here I am particularly interested in model pluralism, for this form of pluralism has also been advocated by those with a more sanguine view of the return of the group, such as the philosopher Kim Sterelny and the biologists Lee Alan Dugatkin and Hudson K. Reeve. Model pluralism has also been elegantly articulated and defended more recently by Benjamin Kerr and Peter Godfrey-Smith. According to these authors, there is no fact of the matter as to which of two (or more) putative vehicles or interactors is the level at which selection operates. In particular, descriptions cast in terms of groups and group selection are equivalent to those cast in terms of individuals and what Dugatkin and Reeve call *broad-sense individualism,* the view that "most evolution arises from selfish reproductive competition among individuals within a breeding population."[4]

Model pluralism has proven popular recently amongst biologists working on the levels of selection, the social insects and the origins of eusociality, and the evolution of multicellular and social life from simpler forms. In general terms, these theorists adopt the view that models positing higher-level processes, such as group selection, do not differ significantly or fundamentally from models positing lower-level processes, such as genic or kin selection. As the entomologists Andrew Bourke and Nigel Franks say in summarizing a chapter devoted to this topic, " . . . colony-level, group, individual, and kin selection are all aspects of gene selection. This means that the practice of attributing traits to, say, either colony-level selection or kin selection is illogical." As the first part of this quotation illustrates, model pluralism is sometimes combined with the denial of agent pluralism in that it suggests that there is a sense in which one of these levels – typically, that of the individual or the gene – is more fundamental than the others.[5]

On Sterelny's construal of broad-sense individualism, groups simply form part of the social environment of individuals. Here the form that individual selection takes is *frequency-dependent* selection, where the fitness of a given trait varies as a function of the frequency of either that trait or some other trait in the group or population. Sterelny suggests that this form of pluralism is applicable to a range of cases, including ant colonies/nests and the elaborate chemical and behavioral warning systems that its members have, kin selection in general, honey bee altruism, and trematode altruism.[6]

Positions similar to model pluralism have been endorsed more generally in debates over the nature of the biological world, particularly when those debates have seemingly either reached a stalement, or when defended positions have led to clashes with widely shared intuitions. For example, faced with the many definitions of biological species, some have concluded that species are "unreal," or that there are simply different species concepts applicable to different explanatory contexts or projects. And to take an example closer to the model pluralism that is my focus here, Richard Dawkins famously replied to the putative departure that his selfish gene theory made from traditional, individual-centered Darwinian theory by suggesting that the two views were simply different but noncompeting perspectives that one could adopt to the agent of selection. Shifting between the gene's-eye view and the individual's-eye view was like undergoing a gestalt shift in looking at a Necker cube. In each case, we have two views each equally anchored in the reality they reflect. This *Neckerphilia*, as we might call it, has been enthusiastically endorsed most recently by Philip Kitcher, and by Ben Kerr and Peter Godfrey-Smith.[7]

As my discussion of species in Part Two perhaps suggests, I am suspicious of these ways of deflating prima facie ontological disagreement. I shall focus my critique on the model pluralism of Lee Dugatkin and Hudson Reeve, and Benjamin Kerr and Peter Godfrey-Smith, and the key argument they give for it. But first I lay out the model pluralist view more fully.

3. A CLOSER LOOK AT MODEL PLURALISM

I have characterized agent pluralism as realist and antifundamentalist about the agents of selection. By contrast, model pluralism is either conventionalist or fundamentalist about these, views that I find problematic. Dugatkin and Reeve express conventionalism in claiming that "individual and trait-group selection are not alternative evolutionary mechanisms; rather, they are alternative pictures of the same underlying mechanism." As Sterelny says in glossing Dugatkin and Reeve's broad-sense individualism, on this view there are only "heuristic differences between the two approaches [broad-sense individualism and group selection theory] . . . the vehicles we should recognize depends on our explanatory and predictive interests."[8]

Central to the arguments of both Dugatkin and Reeve and of Kerr and Godfrey-Smith is the claim that there are mathematical, logical, and explanatory equivalences between broad-sense individualism and multi-level selection theory. These are taken to imply that while one might adopt one or the other of these views of natural selection for pragmatic or heuristic reasons, they do not correspond to alternative processes in nature. Rather, the labels "broad-sense individual selection" and "group selection" pick out one and the same process of natural selection, and function in intertranslatable frameworks. Hence, the differences here between broad-sense individual and group selection perspectives are in the eye of the beholder, not in the reality they purport to describe. As Reeve says elsewhere,

the 'new' group selection models...are not mathematically different from the broad-sense individual selection (e.g., inclusive fitness) models at all. Rather, the group selection models are generated from a fitness-accounting scheme that merely produces an alternative *picture* of the same selective processes described by the inclusive fitness models.[9]

Much the same view is expressed by Sober and Wilson in the passage quoted in the previous section.

Sterelny's form of model pluralism shifts between this type of conventionalism and fundamentalism. Sterelny's view is fundamentalist in that it claims not only that groups can be regarded simply as parts of an individual's environment – and so group selection is recast as frequency-dependent, individual selection – but that this implies that the latter is more fundamental than the former. Such fundamentalism derives from Sterelny's skepticism about the status of trait groups in general as agents of selection. While he holds that trait groups that form superorganisms are (objectively) vehicles, those that are *mere* trait groups, that is, the vast majority of them, are not vehicles and so there can be no level of selection that operates on them. As Sterelny says, "Organisms and superorganisms are real vehicles. There is a fairly objective description of their location in design space. Their existence and location in the biological world is stance-independent. Trait groups that are not cohesive do not share this objective existence as vehicles."[10]

Thus, a translation scheme that allows one to move from descriptions of group selection to those of individual selection should be taken to imply the eliminability of the former in favor of the latter. With the exception of organism-like groups, trait groups do not form a part of the objective fabric of the biological world subject to natural selection. So alternative descriptions of the level at which selection operates, such as that provided by frequency-dependent, individual selection, are to be preferred. This skepticism about the reality of groups as "vehicles" or "interactors" is widespread, particularly amongst biologists who adopt model pluralism, and it introduces a crucial asymmetry between individual (and genic) selection and group selection.

Sterelny thinks that group selection exists only when there are superorganisms, and that this is rarely. In Chapter 8, I expressed my agreement with this latter claim, but I questioned the need for the first. Given the parasitic strategy of argument for defending the integrity of group selection, groups need not be conceived as superorganisms in order to make sense of group selection, since they could be agents of selection without also being manifestors of adaptation or the beneficiaries of selection. If the problems of arbitrariness and ephemerality can be solved short of finding superorganisms in the world, as I have argued they can in the myxoma case, then even if organisms are a paradigm agent of selection, groups need not pass some threshold on a similar metric defined in terms of them in order to be considered agents of selection.

Common to both of these versions of model pluralism is the claim that there are empirically adequate models of evolutionary change that posit different, alternative agents of selection. But to derive the specific forms of model pluralism that I have outlined we need one or the other of two additional assumptions. For Dugatkin and Reeve, this assumption is one about the translatability of claims made in either framework into those of the other; they take this to imply conventionalism about "the" agent or agents of selection. For Sterelny, the missing assumption concerns the unreality of trait groups in general; for him this implies fundamentalism about lower-level agents of selection (perhaps excepting the special case of organism-like groups). Although the shared assumption itself warrants further discussion, I shall concentrate primarily on the translatability assumption and its putative significance for model pluralism. Consider this line of reasoning, as found in Dugatkin and Reeve, and in Kerr and Godfrey-Smith.

Dugatkin and Reeve claim that "the mathematics of the gene-, individual-, kin- and new-group-selection approaches are equivalent," and that "this must be the case and that individual and trait-group selection are not alternative evolutionary mechanisms; rather, they are alternative pictures of the same underlying mechanism." Their claims of the "logical equivalence" and "logical translatability of one picture into the other" are accompanied by an acknowledgement of the heuristic differences between individual and trait-group selection. They take these claims together to entail that claims that pit these two forms of selection against one another are "empirically empty."[11]

Kerr and Godfrey-Smith reinforce this view by showing how the basic parameters in what they call the *individualist* or *contextual* approach and the *multilevel* or *collective* approach to "dealing with group structure in a population" can be defined in terms of one another. This allows them to show the parallels between the fitness structures that each of these models generates and, ultimately, to derive a dynamically sufficient representation of phenotypic change using either model. Since they emphasize the virtues of being able to move between both models, they call their view *gestalt-switching pluralism*. They are Neckerphiliacs.[12]

Underlying this sort of argument is the idea that if alternative models of natural selection can be represented either in a common mathematical framework or by the very same equations, then these models differ at most heuristically, not on some deeper level. Dugatkin and Reeve illustrate this by showing how to translate between equations that state when a trait of an individual will evolve by natural selection, and when an allele will be naturally selected. A shared mathematical structure provides the basis for translating between these different models of selection. As they conclude, "*If broad-sense individual selection, genic selection, and trait-group selection all can be represented by a single condition based only on allele frequencies, then they cannot fundamentally differ from one another.*" This conditional, however, is false, and the more general argument that it reflects fallacious. The sense in which various models and forms of selection can be represented by a shared framework, and the significance of this, need to be explored.[13]

4. REPRESENTATION, TRANSLATION, AND LEVELS

To begin, note that having a shared mathematical framework, or being represented by the very same equations, does not itself entail that two or more processes "cannot fundamentally differ from one another." This is because mathematical models capture just some aspects of the dynamics or kinematics of the processes they model, and they serve as models of those aspects only given further assumptions not represented in the models themselves. Indeed, one should expect at least some mathematical models to be applicable to fundamentally different processes. Two processes represented in the same mathematical framework or by the same mathematical equations can fundamentally differ from one another because the mathematics (1) is an attempt to selectively represent just some aspects of the phenomenon being modeled, and (2) requires further, interpretative assumptions in order to serve as a model of that phenomenon.

Consider some examples in areas other than biology. Decision theory is concerned with how agents ought to make choices, and its formal side provides various rules that constrain such choices, including those specified by utility theory, Bayes's account of probabilistic reasoning, and game theory. Such theories can be applied to a range of agents, including human agents, organizations, insect societies, or anything that can be construed as having preferences and means of attaining them. Applying them in each of these cases will require different auxiliary assumptions, but this "pluralism" implies nothing about whether these various domains of application differ fundamentally or not. The backpropagation algorithm familiar from connectionist cognitive modeling in cognitive science can be used to model not only various cognitive processes (verb acquisition, memory, attention, character recognition), but noncognitive systems that can be construed as networks with basic units whose activation is a function of both past and desired levels of activation. Its application within the cognitive realm does not show that the various cognitive processes "cannot fundamentally differ from one another." Likewise, its application to cognitive and noncognitive domains carries no implication concerning whether these are fundamentally similar or not. Cellular automata have also been used to model cognition as well as the operation of the immune system, and again the existence of shared mathematical frameworks across domains does little to show just how similar those domains are.[14]

This objection – to the general inference suggested by Dugatkin and Reeve and by model pluralists more generally – does not take us very far, however, for two reasons. First, it begs one question at issue by neglecting (or denying) the unifying potential that shared formalisms introduce. Showing how to apply, say, game theory to previously disparate phenomena itself allows those phenomena to be at least partially viewed as instances of the same kind of thing; this itself may constitute a major integrative advance. Moreover, the selective nature of mathematical representations is no barrier to their capturing what is central about a phenomenon. One might point to the "frame shifting" between genic, individual, and group selection at the heart of Sober and Wilson's multilevel view of selection as illustrative of both points.

Second, and more importantly, model pluralists about selection have putatively identified an extensive, fine-grained isomorphism between various models of selection. Model pluralists do not simply claim that the models are result equivalent but that they are causally equivalent in their structures. For example, Dugatkin and Reeve provide a mapping between individualist and trait-group explanations, and Kerr and Godfrey-Smith claim that their "two parameterizations of the fitness structure in the system contain exactly the same information, parceled up differently." Minimally, the onus is on someone who rejects model pluralism to identify differences between the models that cannot reasonably be called "pragmatic" or "heuristic" in nature. I focus in what follows on the parameterizations that Kerr and Godfrey-Smith provide. Their more precise

TABLE 10.1. *A translation scheme for two models of selection*

	Individualist	Multilevel
Definition of fitness structure	α_i = number of copies for an A type in group with i A types β_i = number of copies for a B type in group with i A types	π_i = total number of copies for a group with i A types ϕ_i = number of A copies in a group with i A types / total number of copies in a group with i A types
Relation to other perspective	$\alpha_i = \pi_i\, \phi_i\, /\, \mathrm{I}$ $\beta_i = (1 - \phi_i)\, \pi_i\, /\, n - \mathrm{I}$	$\pi_i = i\, \alpha_i + (n - i)\, \beta_\mathrm{I}$ $\phi_i = i\, \alpha_i\, /\, i\, \alpha_i + (n - i)\, \beta_i$

Source: Portion of Kerr and Godfrey-Smith's Table 1 showing fitness structures of the individualist and multilevel perspectives and their relations. Reproduced with permission of the authors and Kluwer Academic Publisher

characterization of many of the ideas expressed by Dugatkin, Reeve, and Sterelny reveals the depth of the formal equivalence between "individualist" and "multilevel" perspectives on selection and, in my view, makes the strongest case for model pluralism yet provided.[15]

Consider Kerr and Godfrey-Smith's Table 1, part of which is redrawn here (Table 10.1). The fitness structure posited in the individualist perspective is parameterized in terms of the number of copies of the relevant genotypes or phenotypes, α_i and β_i, where the index signifies a relativization to a group with i A types. This reflects a standard individualist view, adapted to explicitly recognize a role for population structure as the *context* in which types occur. The fitness structure in the multilevel perspective, by contrast, is parameterized in such a way as to reflect the group as a *collective* that can serve as the bearer of fitness values, and so captures the idea that natural selection can operate on groups themselves. Thus, π_i and ϕ_i aim to express, respectively, group productivity and the contribution of A types to that productivity. So parameterized, the mathematical equivalences between the two perspectives falls out with some simple algebra.

Suppose that we grant the parameterizations for now. What follows? To derive a conclusion about equivalence in the two perspectives, a conclusion about the translatability between them, we need at least the assumption that these formalizations express the core of those perspectives. One reason to question this, even granting the formalizations, is that the two perspectives differ in their quantificational range. In particular, the individualist perspective quantifies over just individuals while the multilevel perspective quantifies over groups, not only as the bearers of fitness values or as interactors but also as the manifestors of adaptations and the beneficiaries of natural selection. A group may play one or more of these roles in any given case in the multilevel perspective, but none of these roles in the individual perspective. Groups, in short, are agents in the process of natural selection in the multilevel perspective, but are never such in the individualist perspective. This is an ontological difference between the two perspectives, and insofar as it is not reflected in the formalizations, they leave something important out. Moreover, it is difficult to see how adding in parameters to the multilevel model to capture such agentive roles would allow one to preserve the isomorphism between the two formal models, since that would in turn require corresponding parameters in the individualist model. But since individualists do not think that groups play such roles – certainly, not in all cases – this would seemingly require a more radical departure from their perspective than simply relativizing types to population structure.

There is a dilemma here for the model pluralist that we might express in several ways. Consider one of these. If the models are exhaustively constituted by the formal definitions provided in Table 10.1 (and its extensions), then these models leave out something crucial about each of the perspectives they represent. So formal equivalences in the models do not support the claim that the two perspectives are simply different ways of expressing a shared view of the biological world (and so do not compete for the truth about that world). If, on the other hand, we need to extend the formalizations of each perspective to reflect the deeper ontological differences between them, then the onus is on the model pluralist to show how to (a) perform the formalization, and (b) preserve the isomorphism between the two formal models. I have no definitive argument to show that this is impossible, though I have said something briefly about why I suspect that it cannot be done.

Let us return to the parameterization itself. One of its interesting features is that it provides a way to translate formally between the individualist and multilevel views of the bearer of fitness values, interactors. If this is successful, then it captures one of the "ontological" differences between the perspectives – whether groups play this role – itself a significant achievement. But the characterization of both parameters, π and ϕ, should be questioned by proponents of the multilevel perspective, precisely because each ties a group's productivity too closely to the expected numbers of copies of the genotypes or phenotypes in that group. In some circumstances a group's productivity may be measured just by the total

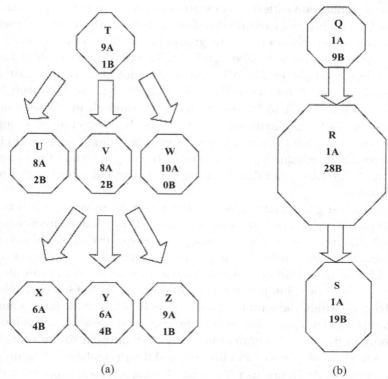

FIGURE 10.1. (a) The T Lineage, containing A and B types. The portion of generation 3 depicted shows how the group structure iterates through group Z. (b) The Q lineage, whose group R in generation 2 exceeds some threshold, leading to a fitness drop. In general, B increases in relative fitness across generations, with the exceptions (from T to W, and from V to Z) mediated by the group structure in the T lineage.

number of copies of individuals it leaves, as the parameterization suggests, but such a measure will be misleading in others where the group structure itself plays a significant role in directing evolutionary change.

For example, consider a group, T, with A and B types that reproduces three groups, U, V, and W, each of its original size, as depicted in Figure 10.1a. At generations 1 and 2, this lineage has precisely the same group productivity (fitness) as defined by π as the lineage in Figure 10.1b. Yet since group size and structure may be ecologically significant, there may be a critical fitness difference between the two lineages, as revealed in generation 3. For example, the T lineage contains Z, which will allow the replication pattern in generation 2 to be reproduced again. And in the Q lineage, R may exceed a threshold, leading to a lower fitness level. Moreover, differences between the groups in Figure 10.1a after the first generation may themselves be significant for evolutionary change. For example, a "pure group" like W breaks the dynamic of decreasing within-group fitness of As relative to Bs. Note that this dynamic is broken in the transition from T to W, and from V to Z, through the "defection" of just one B to A, something that itself may be a function of group structure. π is insensitive to these sorts of causal differences precisely because it conceptualizes group productivity solely in terms of the composition of groups rather than the reproduction and structure of groups themselves.

One might grant that this is true but insignificant because it is the other parameter, ϕ, that is relevant here, for it is here that the underlying causal story is to be told within the multilevel perspective. Again, however, this should be questioned by a proponent of the multilevel perspective, for this parameterization is also cast exclusively in terms of individual offspring or copies. But just as group productivity should sometimes be understood within the multilevel perspective in terms of the production of groups (and not just individuals in those groups), we also need to look beyond an individual's within-group frequency in order to understand the causal interplay between individuals and their population-structured environments. As Figure 10.1 illustrates, having a certain proportion of A types within a group may be necessary for that group to flourish or even survive, and this role may vary across different ecological circumstances. In such a case, the parameterization will offer at best a partial characterization or approximation of the contribution of A types to group productivity.

The basic problem with both parameters is precisely what allows them to serve as the bridge between the individualist and multilevel perspectives: they are themselves too individualistic. Group productivity is parameterized solely in terms of the individual composition of groups and then used to define the particular contributions of individuals to overall group productivity. While this is sometimes an appropriate way to understand groups themselves as bearers of fitness values – since group fitness sometimes just is the sum of the fitness of the individuals it produces – it does not allow one to represent circumstances in which groups play a more agentlike role in natural selection.

One final point about model pluralism. I noted at the outset that model pluralism has proven attractive across the spectrum of views over the levels of selection. But this attraction has a very different basis for proponents of group selection and individualists about selection. David Sloan Wilson, for example, has sought to restore the theory of group selection to respectability over an extended period in which individualistic views have predominated. Thus, he has both distanced his views from "naive group selection theory" of the past and sought to integrate his "new group selection theory" with predominant individualistic views. In this context, model pluralism serves a legitimating function. Individualists, by contrast, are attracted by model pluralism because it allows them to treat new group selection theory as old wine in new bottles. Cases that seem to call out for viewing groups themselves as occupying center stage in the process of natural selection are redescribed from the individualist perspective with groups present but very much in the wings. Phenomena that feature groups themselves as agents of selection exist but are rare, very much the view that George Williams originally defended. In this context, model pluralism functions to maintain the status quo.[16]

Disagreements that lie beneath the surface of the pluralist consensus reflect this contextual difference. One such disagreement concerns agent pluralism and its realist and antifundamentalist standpoint. We have seen that Sober and Wilson endorse both model pluralism and agent pluralism, while individualists remain tempted by fundamentalism and are, at best, neutral or guarded about agent pluralism. One way to read my criticism of the translatability argument for model pluralism is as suggesting that unless it allows us to resolve the ontological issues that agent pluralism raises, it stops short of dissolving the disagreement between individualists and multiselectionists.[17]

5. ENTWINED LEVELS OF SELECTION

Model pluralism is an enticing position for making sense of the debate over the agents of selection, a temptation harder to resist in light of the appeals to the mathematical equivalences between various models of selection. But I have argued that these equivalences do not have the significance that model pluralists have ascribed to them. And I view model pluralism itself as a mistaken way to spell out the intuition that there may not be a determinate answer to the question of just which level is "the" level at which selection occurs in any particular case. Agent pluralism shares this intuition but adopts a realist view of whether selection is operating at the genic, individual, or group level (or all three) in any particular case. The agent pluralist's supposition is that the levels at which selection operates are always sufficiently separable for it to make sense to invoke criteria for preferring one or more levels to others, at least in particular cases.

It is this supposition that I challenge in this section. Various levels of selection are often entwined in that while they are reliably coinstantiated, they do not make isolatable, distinct contributions to the ultimate evolutionary currency, fitness. Especially when two or more levels of selection work in the same direction, and act in ways to mutually reinforce one another, as I think is often the case, isolating the contribution of each level involves an artificial partitioning of the causes of the resulting evolutionary change (or stasis) that does not track biological reality. In what follows, I hope to show how to articulate this idea to arrive at a position that departs from both agent and model pluralism while accommodating the central intuition that motivates these views.

I present the argument in two stages: with respect to properties *at a level*, and then with respect to processes or mechanisms *at different levels*. The first stage aims to show that properties at the group level can be entwined, as can those at the individual level. In the second stage, I argue that the very processes that generate these entwined properties across the two levels are themselves entwined. It is this entwinement "across levels" that challenges the very idea that the biological world separates neatly into layers or levels. I develop both stages of the argument through a detailed, idealized example.

Consider a metapopulation consisting of groups of nesting organisms that vary with respect to their invadability by outsiders. Suppose that the least invadable groups have a certain proportion of distinctly larger individuals, and this is because those individuals are particularly adept at blocking holes in the nest with their larger-than-average bodies. These larger individuals come to specialize in nest defense, and there is some division of labor within the nest of which such a specialization forms a part. Since larger individuals require more food resources, too many of these lead to resource-based problems for the group, and so there is an equilibrium number, m, of large individuals for any nest of size, n, to possess. Here it would be natural to say that certain groups are less invadable because they have a certain proportion, m/n, of larger individuals. Suppose also that individuals are able to grow distinctly larger only because they are members of relatively uninvadable groups, since the corresponding division of labor creates time for more individuals to forage. Suppose also, as a result, that uninvadable groups have a higher level of fitness than invadable groups.

In a population with very few distinctly large individuals there will be individual selection for individuals of greater body size: their proportion will increase within the group over time until it reaches the equilibrium ratio of m/n. This is a form of frequency-dependent selection that selects for an intrinsic property of individuals, body size. But since all and only those individuals with greater body size are also those who function as hole blockers, there is also selection for hole blockers in this population. In the environment in which selection actually occurs, in the terminology introduced in Part One, the realization of one of them is also a realization of the other, and this is no accident. Natural selection is sensitive to the ways in which properties are reliably instantiated, and so it is not fine grained enough to distinguish between entwined properties.

Consider now group selection. Because the groups vary with respect to their invadability, this variation correlates with fitness variation, and that variation is heritable (that is, survives intergenerationally), lower levels of invadability are adaptations, and there is a (group) selective pressure for their evolution. Either groups with low levels of invadability will differentially grow in members, or such groups will be ancestors to more groups than those with higher levels of invadability. But it is also surely true that there is group selection for groups with, or that approximate having, m/n large individuals. Being uninvadable and having m/n large individuals are not identical, for there are possible, even nomologically possible, situations in which they come apart. Yet, in the example as I have described it, they are entwined. And as before, natural selection does not – indeed, *cannot* – distinguish between entwined properties.

Furthermore, not only are properties at both the group level and the individual level entwined, but the two corresponding processes or mechanisms, group and individual selection, are also entwined. Individuals with large body size and that play the hole-blocker function are selected within groups, and there is the selection between groups of lower levels of invadability and having m/n large individuals. But the selective processes at different levels here are not independent. They constitute mutually reinforcing forces whose effects, the properties that are selected, are closely and deeply related, with neither individual- nor group-level properties being causally prior to the other. These reliably coinstantiated selective forces do not make isolatable, distinct contributions to fitness. Moreover, given the case as described, these selective processes can be teased apart only in thought or through experimental interventions that alter the very relationships between them.

Perhaps what I am calling entwinement would be more aptly named *fusion*, since its occurrence implies the oneness of what might, in other cases, be separate and separable forces, in much the way that two winds can fuse to create a gale whose effects are not attributable separably to each of the initially (and in other circumstances, continuingly) distinct winds. Vectors can be used to apportion causal responsibility, and this can have useful predictive value, but there is no meaningful way of repartitioning the gale itself into distinct vectors. Likewise, entwinement or fusion of properties or processes implies that it makes no sense to apportion determinate, partial causal responsibility for the resulting evolutionary change. This is true whether we take the properties of having a certain body size or type and being a hole blocker at the individual level, of being uninvadable and having m/n large individuals at the group level, or the levels themselves.

When either properties or processes merely co-occur, they can be teased apart through experimental means, using various control groups and experimental conditions in order to isolate the distinctive contribution that a property or process makes to the direction and size of the force of natural selection. This is precisely what one might think lies at the heart of experimental approaches to determine the level or levels at which selection operates. Indeed, one might think that experimental intervention is needed in order to determine when properties are lawfully related and when they merely co-occur in a wide range of natural circumstances. But such experimental interventions reach a limit when properties or processes are entwined, since in that case the two do not pre-exist as independent, separable parts of the world. Experimentation and artificial selection may show us that some cases of apparent entwinement are merely apparent, but it can no more distinguish entwined properties than can natural selection. When there is entwinement, further empirical investigation reveals further complexities and entanglements, and the discovery of the workings of isolated properties or processes is an artifact of the experimental investigation, an artifice of our imposing on the world an ontology with distinct levels in an idealized biological hierarchy.

An examination of Michael Wade's classic experimental approach to group selection on flour beetles, two species of *Tribolium*, and the work that it has inspired, will illustrate this point. In a series of elegant and often-discussed experiments, Wade and his colleagues selected for groups of beetles with high rates of population increase, pitting these against both randomized control groups and those in which there was only individual selection. Rather than using a method of panmixis across generation times, as had previous models of group selection, Wade used propagules as the basis for new generations in his artificial (group) selection model. The basic finding was that there are conditions under which the effects of group selection are significant, and significantly stronger than those of individual selection. Wade himself selected for the group-level property rate of population increase, but related studies have selected for other group-level traits in *Tribolium*, such as emigration rate, and multilevel traits in other species, such as leaf area in the cress *Arabidopsis thaliana*.[18]

Although each of these studies shows that a certain property can be most effectively selected through group rather than individual selection, even collectively they do not show that there are no other properties being selected through group or individual selection. Charles Goodnight and Lori Stevens have noted that "[a]ll of the traits examined in group selection experiments are influenced by interactions among individuals," and these interactions point to the tight relationships that exist between various properties in this experimental paradigm. For example, population size and emigration rates in *Tribolium* are determined by (adult) cannibalism rates and population density, respectively. Each of these pairs of group-level properties is entwined in the experimental conditions, and thus both are selected through group selection (high population size with low rates of cannibalism; high emigration rates with high population density). But there is also a range of individual-level, dispositional properties, such as fecundity, aggression, and cannibalism. These are not independent of these group-only and multilevel traits, and are selected along with them.[19]

For example, when the rate of cannibalism – a group-level property – is high, there are more individuals manifesting cannibalism. So we could see individual selection as operating on it. This sort of entwinement underpins Sterelny's suggestion that we "reanalyse the evolution of (say) large flour beetle groups as the evolution of a gene for (say) synchronizing breeding, a gene advantageous only in a particular population-structured context." What I am suggesting is that we give up the fundamentalism that I think drives Sterelny's own talk of "reanalysing," but recognize the reality of the metaphysical entwinement that gives rise to it.[20]

Wade's more recent work on "indirect genetic effects" (IGEs) is interestingly positioned vis-à-vis traditional talk of levels of selection and the notion of entwinement that I have introduced. IGEs occur when there are phenotypic or fitness changes that result from changes to the genes of an individual's social partners, rather than simply from that individual's own genes. While phenotypic expression and fitness always depend on the values of environmental variables, IGEs constitute a malleable environment whose effects on selection can be extreme. For example, the effect of plant height on a given plant's fitness will depend, in part, on the heights of its surrounding neighbors, and changes in an individual's genetic environment can very quickly change the fitness values of its phenotypes. Agrawal, Brodie III, and Wade extended the basic models of IGEs to apply to cases in which population structure directs selection, finding that IGEs could be significantly higher once assumptions of panmixis and linear interactions were relaxed. They remind us that the nonadditive effects of individual and group selection had long been noted in the literature on experimental approaches to group selection, and view their own models as providing an account of the interactions between these levels of selection. This is because the models themselves add in terms that explicitly refer to the effects of interactions between the levels of selection.

However, one might also see this work as posing a challenge to the framework that multiselection theory operates within, and as providing conceptual and mathematical tools more readily viewed as applying, not to a layered biological reality, but to one whose "levels" are inherently entwined with one another. Recent work by Wolf, Wade, and Brodie III, which proposes a general way to model a variety of gene-context interactions, might be viewed as a further step in this direction.[21]

The entwinement of properties and processes in the biological world is clearly a matter of degree, a point that we can use to situate the view I am advocating with respect to our two forms of pluralism. When entwinement is extremely strong or "tight," as it was in my idealized example, then model pluralism takes on some appeal, since the modeling tools of the trade, like natural selection itself, will be insensitive to underlying ontological differences. And when entwinement is extremely weak or "loose," agent pluralism will allow us to approximate the contributions of individual-level and group-level properties and agents to the process of evolution by natural selection. Since I think that we have significantly underestimated the extent to which properties in the biological world are entwined, the limitations of agent pluralism are likely to be more severe than its proponents have allowed.

The idea that biological reality is entwined provides a different starting point, however, from both model and agent pluralism in thinking about the nature of selection. For any given organism the evolutionary interest of various agents – genes, individuals, and groups – are inextricably woven together. None of these putative agents floats free of the others in evolutionary space. To take the standard Darwinian agent, individuals typically mediate their reproductive success through their membership in a group or groups, and do so by means of the replication of their genetic material; genes exist and thus reproduce only in individuals, and groups do so only through the reproductive efforts of the individuals that comprise them. There is thus a clustering of replicative, reproductive, and social interests – not always in harmony but inextricably knitted together – that lies somewhere between the merely fortuitous and the lawful. By taking this entwinement to be ontologically basic, we can see the hierarchically layered view shared by pluralists, and so the issue of how these levels are related, as an artifact of just one way of thinking about the metaphysics of science.[22]

6. LEVELS, ENTWINEMENT, AND THE METAPHYSICS OF SCIENCE

In the previous section, I elaborated on the idea of entwinement as a tool for complicating the conception of natural selection as operating at distinct levels that underlies the forms of pluralism discussed in sections 2–4. Given the ubiquity of talk of levels amongst those engaged in various debates over the metaphysics of science, if the notion of entwinement proves useful in characterizing the "levels of selection," it is likely to have broader application. I want to conclude the substantive part of this chapter by raising some general issues about levels and entwinement with this in mind.

The first of these concerns the metaphorical nature of talk of levels in many scientific contexts. Such talk might be thought to be literal, to correspond to a distinctly layered reality, in physics, where the concern is to mark scale differences between objects of different size. Whatever one thinks of this, it is too optimistic as a view of levels speak in other sciences. For example, the common idea that we should understand talk of the "levels of selection" literally in terms of entities nested in a compositional hierarchy (for example, genes, individuals, families, groups) is confused. First, there is no univocal relational expression that applies to the full range of entities that have been postulated as putative objects of selection. "Compose," "are a part of," "physically constitute," and "are realized in" – to take four of the more obvious predicates relating smaller to larger biological entities that are candidate agents of selection – do not truly, literally, and univocally describe the relationship between adjacent pairs in the following series: alleles, genotypes, cells, organs, bodily systems, organisms, kin groups, families, nests, colonies, species. Second, it is much less clear how properties, processes, events, and states of such entities can stand in these sorts of mereological relationships to one another. For example, supposing that groups are physically composed of individuals, in what sense is the process of group selection physically composed of individual selection? At best, the relationship between these processes would seem to be derivative, in some way, from that which holds between the physical objects to which they are "attached." This problem is more severe in the philosophy of mind and the cognitive sciences, where the levels are often simply designated "mental" and "physical" or their variants "psychological" and "neurological."[23]

The second of these issues concerns the relationship between a layered and an entwined metaphysical view of scientific ontology. There seem two distinct ways to think of the relationship. On the one hand, one could view entwinement as an alternative to the layered view of the world. If one thought that talk of levels was misleading, or was itself responsible for creating some of the puzzles that keep philosophers busy (for example, the mind-body problem), then this is likely to be an attractive view of the entwinement metaphor. If the entwinement metaphor is developed with this in mind, then talk of fusion may prove more apt and useful than that of entwinement. On the other hand, one could view entwinement as a supplement to the view of the world as layered, one that either is applicable to special cases and circumstances or that allows one to fine-tune or correct the predominant layered view of scientific ontology. Both of these views may be useful in thinking further about the nature of selection.

The third issue is how we can both make more precise and generalize the notion of entwinement. For, at root, it too gains much of whatever appeal it has to its metaphorical nature. Despite the pioneering work of the philosophers of science Mary Hesse and Richard Boyd, both building on Max Black's views of metaphor, that established a place for metaphor in science, this remains an uncomfortable place for many offering an analytic understanding of scientific processes and debates over them. I have suggested that phenotypic properties are entwined at a level just when they are reliably coinstantiated but do not make distinct contributions to the fitness of their bearer, and that processes across levels are entwined just when *they* are reliably coinstantiated but do not make distinct or independent contributions to the fitness of their respective bearers. There are natural ways to generalize such notions of entwinement to properties and processes more generally, but how plausible these are will turn, in part, on how we conceptualize the relationship between levels and entwinement.[24]

7. THE LURE OF SMALLISM AND THE COGNITIVE METAPHOR

At a number of points in this chapter, smallism – discrimination in favor of the small over the not-so-small – has festered just beneath the skin of the discussion. In this section, I should like to bring this to the surface and offer some reflections on the connections between smallism and both the standard Darwinian view of natural selection and its recent extensions.

Although I have not focused on the idea of selfish genes in this chapter or in Part Four more generally, it is clear that much of the motivation for the idea, and perhaps its popularity, reveals a smallist bias. This bias is reflected in the claim that we should think about natural selection in terms of high-fidelity replicators, for smaller entities will replicate with higher fidelity than the larger entities they compose. It is also manifest in Dawkins's very distinction between replicators and vehicles, which suggests the small as the locus of evolutionary activity, as the agents in the driver's seat of evolutionary change, with the large simply being taken for a ride.

Particularly interesting about the way in which smallism is played out within the theory of genic selection is that genes not only take on the functional role of individuals as the agents of selection: they also come to appropriate the very characteristics distinctive of individual organisms like us. They act for the sake of their own interest understood in terms of their replicative success, and so are seen as adopting strategies that allow this interest to be served. They are thus engaged in the sort of means-end reasoning that has been used to characterize rational agency at least since Aristotle: they have a goal that is set, and ways to achieve that goal.

Of course, no one really thinks that genes have the sort of character sets that human agents have, or are rational in the same sense that we are. But we often talk in ways that readily create the impression that genes are little people. The fact that this sort of attribution of paradigmatic rationalistic character sets is mere "as if" attribution points, I think, to a relationship between smallism and the cognitive metaphor. In Part One, I claimed that smallism lurked in the background of the various forms of individualism that I discussed there. Smallism takes us from individuals to genes as agents of selection, but genes then are given the very characteristics of individuals, even if only in a metaphorical form. Thus, there is an irony in the recent proposed shift in attention within the literature on the levels of selection to the question of "how natural selection among lower-level biological vehicles creates higher-level vehicles." These lower-level vehicles are already imbued with the very characteristics whose existence is to be explained by the resulting smallist narrative.[25]

This interplay between smallism and the cognitive metaphor also exists within the standard Darwinian view of natural selection. Individuals are construed as agents enacting strategies of various levels of complexity. At least in cases in which the organisms themselves are thought to be psychologically relatively simple, this strategic complexity clearly invokes a cognitive metaphor, and does not aim to depict an underlying cognitive reality. Since the dynamics within and between populations – the changes, trends, and stabilities in their genic and genotypic frequencies – are explained in terms of the phenotypic strategies employed by individuals, even the standard Darwinian view adopts a smallist view of evolutionary change, one that prioritizes the activity of individuals in explaining the properties of the populations that such individuals form.

By contrast, the variation on pluralism about the agents of selection that admits the phenomenon of entwinement that I defended in the preceding section departs from smallism. It does so both in resisting the temptation to reduce or explain the large in terms of the small, and in suggesting that individuals should be conceived not only in terms of their constituents but also in terms of what they in turn constitute.

This perspective on evolutionary biology parallels the perspective on cognition that I have advocated in *Boundaries of the Mind* in defending an externalist view of psychology. There I said that externalist psychology could take two forms – taxonomic and locational – and I argued that especially the latter of these pointed to the limitations of both individualism about the mind and the smallist metaphysics on which that view rests. Although these views share the idea that the individual needs to be transcended in order to arrive at an ontology that does justice to the richness of the corresponding domain, pluralism about the agents of selection and externalism about cognition do so in their own ways. On the externalist view of cognition that I articulated, the individual cognizer is the subject of cognition, and the agent remains a locus of control, however wide that individual's psychology becomes. With pluralism about the agents of selection, by contrast, there are cases in which the individual organism, while not quite invisible to the process of natural selection, takes on a secondary role as the entities on which selection acts are either larger or smaller than the individual.

8. BEYOND METAPHOR'S GRASP

Model pluralists are right that there is, in at least some cases, no fact of the matter as to whether selection operates at the (broad-sense) individual level or at the group level. But this is no cause for us to give in to the temptations of Neckerphilia. The absence of a fact of the matter is not due to the intertranslatability of individual and group-level models, nor because one is reducible to the other. Rather, it is a result of the fact that the reality that these models describe includes entwined or fused properties and processes. Thus, the resulting indeterminacy as to just which property is selected, or just which process governs its selection, is not a legacy of our epistemic limitations. Instead, it is a reflection of a biological reality that does not always fall neatly into distinct levels, as proponents of agent pluralism imply. If this introduction of entwinement stimulates discussion in this and other areas in which philosophers of science and others have complacently appealed to the metaphor of levels, so much the better.

Notes

Chapter 1

1. I shall have more to say about Aristotle later in this chapter, and take a closer look at Spencer in Chapter 3. For Fritjof Capra's views, see his *The Hidden Connections: Integrating the Biological, Cognitive, and Social Dimensions of Life into a Science of Sustainability* (New York: Doubleday, 2002). See page 37 for both the quotation and the information on the Indo-European languages. Although etymology provides support for a relationship between breath and life, as Capra says, the Indo-European languages he relies on manifest no such link between life and mind. Thanks to Selina Stewart for this point.

2. For some discussion of particular aspects of this Cartesian legacy, see my *Cartesian Psychology and Physical Minds: Individualism and the Sciences of the Mind* (New York: Cambridge University Press, 1995); and Andy Clark, *Being There: Putting Brain, Body, and World Together Again* (Cambridge, MA: MIT Press, 1997).

3. Todd A. Grantham, "Hierarchical Approaches to Macroevolution: Recent Work on Species Selection and the 'Effect Hypothesis,'" *Annual Review of Ecology and Systematics*, 26 (1995), pages 301–321.

4. Table 1.1 is reproduced from Eldredge's Table 6.3 in his *Unfinished Synthesis* (New York: Oxford University Press, 1985), page 166, which in turn is a variant on a table from a coauthored paper with Salthe, "Hierarchy and Evolution," *Oxford Surveys in Evolutionary Biology*, 1 (1984), pages 182–206.

5. See Scott Atran, "Itzaj Maya Folkbiological Taxonomy: Cognitive Universals and Cultural Particulars," in Doug Medin and Scott Atran (editors), *Folkbiology* (Cambridge, MA: MIT Press, 1999). The quotation is from page 120 and "specieslike" is Atran's coinage.

6. For Aristotle's views on individual, species, and natural kinds in his *Categories*, see Chapter 5, especially 2a11–4b. The quotation I provide can be found at 2b7–8. See *Aristotle's Categories and De Interpretatione*. Translated by J. L. Ackrill (Oxford: Oxford University Press, 1963).

7. For the classic challenges to essentialism in the philosophy of biology, see David Hull, "The Effect of Essentialism on Taxonomy: Two Thousand Years of Stasis," *British Journal for the Philosophy of Science*, 15 (1965), pages 314–326 and 16 (1965), pages 1–18. Reprinted in Marc Ereshefsky (editor), *The Units of Evolution* (Cambridge, MA: MIT Press, 1992); and Ernst Mayr, *The Growth of Biological Thought* (Cambridge, MA: Harvard University Press, 1982), especially Chapters 2, 8, and 9. These views of the history of essentialism have recently come under scrutiny from historians of biology. See, for example, Mary P. Winsor, "Non-essentialist Methods in Pre-Darwinian Taxonomy," *Biology and Philosophy* 18 (2003), pages 387–400; Paul Thompson, "'Organization,' 'Population,' and Mayr's Rejection of Essentialism in Biology," in D. Sfendoi-Mentzou, J. Hattiangadi, and D. M. Johnson (editors), *Aristotle and Contemporary Science, Volume 2* (New York: Peter Lang, 2001), pages 173–183; and Gordon R. McOuat, "Species, Rules, and Meaning: The Politics of Language and the Ends of Definitions in 19th Century Natural History," *Studies in the History and Philosophy of Science* 27 (1996), pages 473–519.

8. For pluralism in the philosophy of biology, see John Dupré, *The Disorder of Things: Metaphysical Foundations of the Disunity of Science* (Cambridge, MA: Harvard University Press, 1993), especially Chapters 1–3; and Philip Kitcher, "Species" (1984), and "Function and Design" (1993), both reprinted in his recent collection, *In Mendel's Mirror: Philosophical Reflections on Biology* (New York: Oxford University Press, 2003). The quotation from Kitcher comes from page xi of his preface to this volume.

9. The quotation from Dupré is from his "On the Impossibility of a Monistic Account of Species," in Robert A. Wilson (editor), *Species: New Interdisciplinary Essays* (Cambridge, MA: MIT Press, 1999), at page 16.

10. See the works cited in note 8 above for views of reductionism, as well as Dupré's *Human Nature and the Limits of Science* (New York: Oxford University Press, 2001) for evolutionary psychology, and Kitcher's, *The Lives to Come: The Genetical Revolution and Human Possibilities* (New York: Touchstone Books, 1996) for eugenics. In addition, in Kitcher's *In Mendel's Mirror*, see especially the two essays "Race, Ethnicity, Biology, Culture" (1999) and "Battling the Undead: How (and How Not) to Resist Genetic Determinism" (2001) for a sample of Kitcher's more recent views of biology and society.

11. Andrew Bourke and Nigel Franks, *Social Evolution in Ants* (Princeton, NJ: Princeton University Press, 1995), page 67.

12. Dobzhansky's view is expressed in his "Nothing in Biology Makes Sense Except in the Light of Evolution," *American Biology Teacher* 35 (1973), pages 125–129. For more on the Modern Synthesis, see Ernst Mayr and William B. Provine (editors), *The Evolutionary Synthesis: Perspectives on the Unification of Biology* (Cambridge, MA: Harvard University Press, 1980); and Vassiliki Betty Smocovitis, *Unifying Biology: The Evolutionary Synthesis and Evolutionary Biology* (Princeton, NJ: Princeton University Press, 1996).

13. See Richard Dawkins, *The Selfish Gene* (New York: Oxford University Press, 2nd edition, 1989), page 21.

14. For the classic statement of the gene's-eye view and its application to altruism, see William Hamilton, "The Genetical Evolution of Social Behaviour I,"

Journal of Theoretical Biology, 7 (1964), pages 1–16, and "The Genetical Evolution of Social Behaviour II," *Journal of Theoretical Biology*, 7 (1964), pages 17–52. For a recent view of altruism within the framework of multilevel selection that recasts the relationship between genic and group selection, see Elliott Sober and David Sloan Wilson, *Unto Others: The Psychology and Evolution of Unselfish Behavior* (Cambridge, MA: Harvard University Press, 1998).

15. Stephen Jay Gould's *The Structure of Evolutionary Theory* (Cambridge, MA: Harvard University Press, 2002) constitutes the most extensive development of this extended Darwinian view, taking species and clades to be primary agents of selection for macroevolution.

16. On niche construction, see Kevin N. Laland, F. John Odling-Smee, and Marcus W. Feldman, "Niche Construction, Biological Evolution, and Cultural Change," *Behavioral and Brain Sciences*, 23 (2000), pages 131–175; and F. John Odling-Smee, Kevin N. Laland, and Marcus W. Feldman, *Niche Construction: The Neglected Process in Evolution* (Princeton, NJ: Princeton University Press, 2003). On community genetics, see the special editorial feature compiled by Anurag Agrawal in *Ecology*, 84 (2003), pages 545–601. This feature contains lead articles by Claudia Neuhauser et al. and Thomas Whitham et al., together with eight commentaries.

17. For a recent, detailed treatment of the organism as artifact analogy and the many issues that it impinges on, see Tim Lewens, *Organisms and Artifacts* (Cambridge, MA: MIT Press, 2004). For the origins of game theory, see John von Neumann and Oscar Morgenstern, *The Theory of Games and Economic Behaviour* (Princeton, NJ: Princeton University Press, 2nd edition, 1953), originally published in 1944.

18. For an excellent, general work on the human sciences, with a focus on psychology, see Roger Smith, *The Norton History of the Human Sciences* (New York: Norton, 1997).

Chapter 2

1. For discussion of a related debate between internalists and externalists in biology, see Peter Godfrey-Smith, *Complexity and the Function of Mind in Nature* (New York: Cambridge University Press, 1996), especially Chapter 2.

2. On homuncular functionalism, see Robert Cummins, *The Nature of Psychological Explanation* (Cambridge, MA: MIT Press, 1983); Daniel C. Dennett, "Artificial Intelligence as Philosophy and as Psychology," reprinted in his *Brainstorms: Philosophical Essays on Mind and Psychology* (Cambridge, MA: MIT Press, 1978); and William G. Lycan, *Consciousness* (Cambridge, MA: MIT Press, 1987), especially Chapter 4.

3. On multiple realizability, see Jaegwon Kim, "The Myth of Nonreductive Materialism" (1989), and "Multiple Realization and the Metaphysics of Realization" (1992), both reprinted in his *Supervenience and Mind: Selected Philosophical Essays* (New York: Cambridge University Press, 1993); and Lawrence Shapiro, "Multiple Realizations," *Journal of Philosophy*, 97 (2000), pages 635–654, and *The Mind Incarnate* (Cambridge, MA: MIT Press, 2004).

4. See François Jacob, *The Logic of Life* (Princeton, NJ: Princeton University Press, 1973). The quotations are taken from pages 9 and 16.
5. *The Logic of Life*, pages 1–2.
6. Jacob emphasizes the importance of the cell theory for accounts of hereditary transmission in *The Logic of Life*, pages 111–129. The quotation from E. B. Wilson comes from *The Cell in Development and Inheritance* (London: Macmillan, 1900), page 1.
7. See William Bechtel and Robert Richardson, *Discovering Complexity: Decomposition and Localization as Strategies in Scientific Research* (Princeton, NJ: Princeton University Press, 1993). The quotation is taken from page 116.
8. See George C. Williams, "A Defense of Reductionism in Evolutionary Biology," in Richard Dawkins and Mark Ridley (editors), *Oxford Surveys in Evolutionary Biology, Volume 2* (Oxford: Oxford University Press, 1985), page 24. For Williams's earlier views, see his *Adaptation and Natural Selection: A Critique of Some Current Evolutionary Thought* (Princeton, NJ: Princeton University Press, 1966). For his later, somewhat more liberal views about the agents of selection, see his *Natural Selection: Domains, Levels, and Challenges* (New York: Oxford University Press, 1992).
9. Richard C. Lewontin, *The Triple Helix: Gene, Organism, and Environment* (Cambridge, MA: Harvard University Press, 2000), page 3.
10. For discussion of the immune system and the self, see Alfred Tauber, *The Immune Self: Theory or Metaphor?* (New York: Cambridge University Press, 1994).
11. David Sloan Wilson, "Altruism and Organism: Disentangling the Themes of Multilevel Selection Theory," *American Naturalist* 150 (1997), supplement, pages S122–S134. The quotation is taken from page S128.
12. See Mary Douglas, *How Institutions Think* (Syracuse, NY: Syracuse University Press, 1986), page x of the Preface.

Chapter 3

1. These examples came to my attention via the following discussions. The humungous fungus: Jack Wilson, *Biological Individuality: The Identity and Persistence of Living Entities* (New York: Cambridge University Press, 1999), pages 23–25. The mother-fetus-placenta: Sarah Blaffer Hrdy, *Mother Nature: A History of Mothers, Infants, and Natural Selection* (New York: Pantheon, 1999), Chapter 18. Coral reefs: J. Scott Turner, *The Extended Organism: The Physiology of Animal – Built Structures* (Cambridge, MA: Harvard University Press, 2000), Chapter 2.
2. For further discussion of this example, see not only Jack Wilson, *Biological Individuality*, pages 23–25, but also the original discussion of Myron L. Smith, Johann N. Bruhn, and James B. Anderson, "The Fungus *Armillaria bulbosa* is Among the Largest and Oldest Living Organisms," *Nature*, 356 (1992), pages 428–433; and the subsequent discussions by Clive Brasier, "A Champion Thallus," *Nature*, 356 (1992), pages 382–383, and by Kathy Svitil, *Discover*, 69 (1993), pages 69–70. The quotation is taken from the paper by Smith, Bruhn, and Anderson, on page 431. For recent reports of even more humungous fungi of the same species, search on the web using the keyword "humungous fungus."

3. Apart from Hrdy, *Mother Nature*, Chapter 18, see also David Haig, "Genetic Conflicts of Human Pregnancy," *Quarterly Review of Biology*, 68 (1993), pages 495–532; and Mark Ridley, *Mendel's Demon: Gene Justice and the Complexity of Life* (London: Phoenix, 2001), Chapter 8. For Robert Trivers's pioneering work on parent-offspring conflict, see his "Parent-Offspring Conflict," *American Zoologist*, 14 (1974), pages 249–264.

4. For further discussion, see J. Scott Turner, *The Extended Organism*, Chapters 2 and 5. The quotations are taken from page 24 and from page ix of the Preface, respectively.

5. For a recent attempt to define life through an elaborate conceptual analysis that is aimed at a range of puzzle cases, ignores others, and leans heavily on certain intuitions, see Gary S. Rosencrantz, "What is Life?," in Tian Yu Cao (editor), *Proceedings of the World Congress of Philosophy*, Volume 10 (Bowling Green: Philosophy Documentation Center, 2001), pages 125–134.

6. For an example of the sort of Quinean view I have in mind, see Paul M. Churchland, *Scientific Realism and the Plasticity of Mind* (New York: Cambridge University Press, 1979). For more general discussion of naturalism and philosophical analysis, see Philip Kitcher, "The Naturalists Return," *Philosophical Review*, 101 (1992), pages 53–114.

7. See Walter Elsasser, *Atom and Organism* (Princeton, NJ: Princeton University Press, 1965), on the special role of something like intrinsic heterogeneity in living systems. On the Gaia Hypothesis, see James Lovelock, *Gaia: A New Look at Life on Earth* (New York: Oxford University Press, 1995).

8. John Harper, *Population Biology of Plants* (New York: Academic Press, 1977). For further discussion, see Richard Dawkins, *The Extended Phenotype* (Oxford: Oxford University Press, 1982), Chapter 14. Jack Wilson's views are developed in his *Biological Individuality: The Identity and Persistence of Living Entities* (New York: Cambridge University Press, 1999), especially Chapters 3 and 5.

9. For the classic statement of the unity of science hypothesis, and discussion of some of its limitations, see Paul Oppenheim and Hilary Putnam, "Unity of Science as a Working Hypothesis," in Herbert Feigl, Michael Scriven, and Grover Maxwell (editors), *Minnesota Studies in the Philosophy of Science, Volume II: Concepts, Theories, and the Mind-Body Problem* (Minneapolis: University of Minnesota Press, 1958), pages 3–36.

10. See Richard N. Boyd, "Realism, Anti-foundationalism, and the Enthusiasm for Natural Kinds," *Philosophical Studies*, 61 (1991), pages 127–148; "Homeostasis, Species, and Higher Taxa," in Robert A. Wilson (editor), *Species: New Interdisciplinary Essays* (Cambridge, MA: MIT Press, 1999), pages 141–185; and "Kinds, Complexity, and Multiple Realization," *Philosophical Studies*, 95 (1999), pages 67–98. See also Hilary Kornblith, *Inductive Inference and its Natural Ground* (Cambridge, MA: MIT Press, 1993); and Ruth Garrett Millikan, *On Clear and Confused Ideas: An Essay About Substance Concepts* (New York: Cambridge University Press, 2000), Chapter 2.

11. For three essays on the HPC view of species, see Richard Boyd, "Homeostasis, Species, and Higher Taxa"; Paul Griffiths, "Squaring the Circle? Natural Kinds with Historical Essences"; and Robert A. Wilson, "Realism, Essence,

and Kind: Resuscitating Species Essentialism?," all in Robert A. Wilson (editor), *Species: New Interdisciplinary Essays* (Cambridge, MA: MIT Press, 1999), pages 141–228. On the role of essentialism in Darwinian biology, see Elliott Sober, "Evolution, Population Thinking, and Essentialism," *Philosophy of Science*, 47 (1980), pages 350–383.

12. The idea that bodily organs form lineages forms a key part of the distinctively biological view of thought of Ruth Garrett Millikan, in both her *Language, Thought, and Other Biological Categories* (Cambridge, MIT Press, 1984) and in her essays since then, many of the most important of which are collected in her *White Queen Psychology and Other Essays for Alice* (Cambridge, MIT Press, 1993).

13. See James Griesemer, "Reproduction and the Reduction of Genetics," in Peter Beurton, Raphael Falk, and Hans-Jörg Rheinberger (editors), *The Concept of the Gene in Development and Evolution: Historical and Epistemological Perspectives* (New York: Cambridge University Press, 2000), pages 240–285. See also Stephen Jay Gould's discussion of what he calls *plurifaction* in his *The Structure of Evolutionary Theory* (Cambridge, MA: Harvard University Press, 2002), especially page 611.

14. On the need for a view of life cycles and development that is appropriately broad, see Leo Buss, *The Evolution of Individuality* (Princeton, NJ: Princeton University Press, 1987), pages 20–22.

15. The symbiotic theory of organelle development is due to Lynne Margulis. See, for example, her *Symbiosis in Cell Evolution* (New York: W. H. Freeman, 1993).

16. See John Tyler Bonner, *Life Cycles* (Princeton, NJ: Princeton University Press, 1993), page 15; and Paul Griffiths and Russell Gray, "Developmental Systems and Evolutionary Explanation," *Journal of Philosophy*, 91 (1994), pages 277–304. Griffiths and Gray identify organisms with life cycles on page 292.

17. See *Life Cycles*, pages 17–18, for Bonner's four stages.

18. See Nancy Moran and Paul Baumann, "Phylogenetics of Cytoplasmically Inherited Microorganisms of Arthropods," *Trends in Ecology and Evolution*, 9 (1994), pages 15–20. The example is discussed in brief by Paul Griffiths and Russell Gray, "Darwinism and Developmental Systems," in Susan Oyama, Paul E. Griffiths, and Russell D. Gray (editors), *Cycles of Contingency: Developmental Systems and Evolution* (Cambridge, MA: MIT Press, 2001), pages 195–218.

19. See Herbert Spencer, *Principles of Biology, Volume I* (New York: D. Appleton, 1898, revised and enlarged edition), originally published in 1866. The quotations can be found on pages 249–250. Peter Godfrey-Smith discusses Spencer's views of biology more generally as exemplifying a version of what he calls the environmental complexity thesis in his *Complexity and the Function of Mind in Nature* (New York: Cambridge University Press, 1996), Chapter 3.

20. The earlier paper to which I refer here is Herbert Spencer, "A Theory of Population, Deduced from the General Law of Animal Fertility," *Westminster Review*, 1 (1852), pages 468–501.

21. For the quotation, see *Principles of Biology*, page 85.

22. For the restatement, see *Principles of Biology*, page 99.

23. The quotation is from the *Principles of Biology*, page 81. On the "substance of the mind," see Spencer's *Principles of Psychology: Volume I* (New York: Appleton, 1866), Part II, Chapter 1.

24. Julian Huxley, *The Individual in the Animal Kingdom* (London: Cambridge University Press, 1912), page 28.

25. For Huxley on heterogeneity and unity, see *The Individual in the Animal Kingdom*, pages 10–12.

26. For some discussion of organisms as wholes, and their contrast with other material objects, see Peter van Inwagen, *Material Beings* (Ithaca, NY: Cornell University Press, 1990), especially section 9 and forward.

27. Huxley, *The Individual in the Animal Kingdom*, page 25.

Chapter 4

1. For Dennett's view of the intentional stance, see his collection of essays *The Intentional Stance* (Cambridge, MA: MIT Press, 1987), where he in part addresses issues of realism about the intentional stance. These are taken up further in his "Real Patterns," *Journal of Philosophy*, 87 (1991), pages 27–51.

2. Mackie's views, including his discussion of the example of the disgustingness of a fungus, are first expressed in his "A Refutation of Morals," *Australasian Journal of Philosophy*, 24 (1946), pages 77–90. But his best known development of these views are contained in his *Ethics: Inventing Right and Wrong* (Harmondsworth, Middlesex: Penguin, 1977). The projectivism outlined there has its most sophisticated defense in Simon Blackburn's quasi-realism, see his *Essays in Quasi-Realism* (New York: Oxford University Press, 1993).

3. Questions of the eliminability of the cognitive metaphor deserve further (but separate) consideration. On the eliminability of intentional and teleological descriptions within immunology, for example, see Mohan Matthen and Ed Levy, "Teleology, Error, and the Human Immune System," *Journal of Philosophy*, 81 (1984), pages 351–372; and Peter Melander, "How Not to Explain the Errors of the Immune System," *Philosophy of Science*, 60 (1993), pages 223–241.

4. On supernatural agents from the perspective of psychologists, see James L. Barrett and Frank C. Keil, "Conceptualizing a Non-Natural Entity: Anthropomorphism in God Concepts," *Cognitive Psychology*, 31 (1996), pages 219–247; and Jesse M. Bering, "Intuitive Conceptions of Dead Agents' Minds: The Natural Foundations of Afterlife Beliefs as Phenomenological Boundary," *Journal of Cognition and Culture*, 2 (2002), pages 263–308.

5. See Albert Michotte, *The Perception of Causality* (Andover, MD: Metheun, 1963); and F. Heider and M. Simmel, "An Experimental Study of Apparent Behavior," *American Journal of Psychology*, 57 (1944), pages 243–259.

6. For a variety of perspectives on causal cognition and its relationship to psychological and biological agency, see the essays in Dan Sperber, David Premack, and Ann Premack (editors), *Causal Cognition: A Multidisciplinary Debate* (New York: Oxford University Press, 1995).

7. For Gustav Le Bon's views, see especially his *The Crowd: A Study of the Popular Mind* (Dunwoody, GA: Norman S. Berg, 1968, 2nd edition), originally

published in 1895. For David Sloan Wilson's articulation of the group mind hypothesis in contemporary biology, see especially his "Incorporating Group Selection into the Adaptationist Program: A Case Study Involving Human Decision Making," in J. Simpson and D. Kendrick (editors), *Evolutionary Social Psychology* (Hillsdale, NJ: Erlbaum, 1997).

8. The analogy between organs and organisms is used, for example, by David Sloan Wilson as part of his multilevel theory of selection. See, for example, "Altruism and Organism: Disentangling the Themes of Multilevel Selection Theory," *American Naturalist*, 150 (1997), pages S122–S134. See also David Seeley, *The Wisdom of the Hive: The Social Psychology of Honey Bee Colonies* (Cambridge, MA: Harvard University Press, 1995).

9. See, for example, Steven M. Stanley, *Macroevolution: Pattern and Process* (San Francisco: W. H. Freeman,1979), Chapter 7; and Stephen Jay Gould, *The Structure of Evolutionary Theory* (Cambridge, MA: Harvard University Press, 2002), Chapter 8, especially pages 714–744.

10. For representative and classic work of the Chicago School, see W. C. Allee, A. E. Emerson, O. Park, T. Park, and K. P. Schmidt, *Principles of Animal Ecology* (Philadelphia, PA: W. B. Saunders, 1949). On populational physiology, see Thomas Park, "Studies in Population Physiology: The Relation of Numbers to Initial Population Growth in the Flour Beetle *Tribolium Confusum* Duval," *Ecology*, 13 (1932), pages 172–181. And on superorganisms, Alfred Emerson, "Social Coordination and the Superorganism," *American Midland Naturalist*, 21 (1939), pages 182–209; "Basic Comparisons of Human and Insect Societies," *Biological Symposia*, 8 (1942), pages 163–176; and "The Biological Basis of Social Cooperation," *Illinois Academy of Science Transactions*, 29 (1946), pages 9–18.

11. J. Scott Turner, *The Extended Organism: The Physiology of Animal-Built Structures* (Cambridge, MA: Harvard University Press, 2000).

12. For more detailed discussion of locational externalism, see my *Boundaries of the Mind* (New York: Cambridge University Press, 2004), especially Chapter 7.

13. See Leo Buss, *The Evolution of Individuality* (Princeton, NJ: Princeton University Press, 1987); John Maynard Smith and Eörs Szathmáry, *The Major Transitions in Evolution* (New York: Oxford University Press, 1995); and Richard Michod, *Darwinian Dynamics: Evolutionary Transitions in Fitness and Individuality* (Princeton, NJ: Princeton University Press, 1999). The quoted phrase comes from page 3 of Michod's book.

14. *Major Transitions*, page 18.

15. See Richard Dawkins, *The Extended Phenotype: The Gene as the Unit of Selection* (Oxford: Oxford University Press, 1982). The longer quotation is taken from page 264, and the shorter one from page 252.

16. See J. William Schopf, *Cradle of Life: The Discovery of Earth's Earliest Fossils* (Princeton, NJ: Princeton University Press, 1999). For some discussion, see David L. Nanney and Robert A. Wilson, "Life's Early Years," *Biology and Philosophy*, 16 (2001), pages 735–748.

17. For Woese's three domains proposal, see Carl R. Woese, Otto Kandler, and Mark L. Wheeler, "Towards a Natural System of Organisms: Proposal for the

Domains Archaea, Bacteria, and Eucarya," *Proceedings of the National Academy of Sciences USA*, 87 (1990), pages 4576–4579. For his more recent views of the universal ancestor, see his "The Universal Ancestor," *Proceedings of the National Academy of Sciences USA*, 95 (1998), pages 6854–6859; and "Interpreting the Universal Phylogenetic Tree," *Proceedings of the National Academy of Sciences USA*, 97 (2000), pages 8392–8396. The quotation is from the 1998 paper, at page 6856.

18. Chris Langton, "Studying Life With Cellular Automata," *Physica D*, 22 (1986), pages 120–149. The quotation is on page 147. See also his "Artificial Life" in his edited volume *Artificial Life: Santa Fe Studies in the Sciences of Complexity 6* (Reading, MA: Addison-Wesley, 1989).

19. For the quotation from Thomas Ray, see his "An Approach to the Synthesis of Life," in Chris Langton et al. (editors), *Artificial Life II, Sante Fe Studies in the Sciences of Complexity 10* (Redwood City, CA: Addison-Wesley, 1991), page 372. That from Farmer and Belin is taken from the abstract to their "Artificial Life: The Coming Revolution" in the same volume.

20. See Richard Dawkins, "The Evolution of Evolvability," in Chris Langton (editor), *Artificial Life*; Daniel Hillis, "Co-evolving Parasites Improve Simulated Evolution as an Optimization Procedure," in Chris Langton et al. (editors), *Artificial Life II*; and Thomas Ray, "An Approach to the Synthesis of Life."

21. Here I have adapted Daniel Dennett's classic discussion of original and derived intentionality in his "Evolution, Error, and Intentionality," reprinted in his *The Intentional Stance* (Cambridge, MA: MIT Press, 1987), pages 287–321.

Chapter 5

1. The quotation is from Scott Atran, "Itzaj Maya Folkbiological Taxonomy: Cognitive Universals and Cultural Particulars," in Doug Medin and Scott Atran (editors), *Folkbiology* (Cambridge, MA: MIT Press, 1999), page 120. Again, "specieslike" is Atran's coinage.

2. For the quotation from Mishler and Donoghue, see their "Species Concepts: The Case for Pluralism," *Systematic Zoology*, 31 (1982), pages 491–503, reprinted in Marc Ereshefsky, *The Units of Evolution: Essays on the Nature of Species* (Cambridge, MA: MIT Press, 1992), at page 131. For that of Dupré, see his *The Disorder of Things: Metaphysical Foundations for the Disunity of Science* (Cambridge, MA: Harvard University Press, 1993), at page 57. And for Kitcher, see his "Species," *Philosophy of Science*, 51 (1984), pages 308–333, also reprinted in *The Units of Evolution*; the quote is on page 317 of the reprint.

3. As a self-conscious school of taxonomy, pheneticism came to the fore through the numerical taxonomy of Robert R. Sokal and Peter H. A. Sneath, *Principles of Numerical Taxonomy* (San Francisco: Freeman, 1963). See also Robert R. Sokal and Theodore J. Crovello, "The Biological Species Concept: A Critical Evaluation," *American Naturalist*, 104 (1970), pages 12–153, reprinted in *The Units of Evolution*.

4. For statements of these views, see Ernst Mayr, *The Growth of Biological Thought: Diversity, Evolution, and Inheritance* (Cambridge, MA: Harvard University Press, 1982), pages 283–295 on the biological species concept; Hugh H. Paterson, "The Recognition Concept of Species," in Elisabeth S. Vrba (editor), *Species and Speciation* (Pretoria: Transvaal Museum Monograph No. 4); and Alan Templeton, "The Meaning of Species and Speciation: A Genetic Perspective," in Daniel Otte and John A. Endler (editors), *Speciation and its Consequences* (Sunderland, MA: Sinauer, 1989) for variations on this concept. On genealogical views, see Joel Cracraft, "Species Concepts and Speciation Analysis," in *Current Ornithology*, 1 (1983), pages 159–187, and Ed Wiley, "The Evolutionary Species Concept Reconsidered," *Systematic Zoology*, 27 (1978), pages 17–26. All of these papers are conveniently reprinted in Ereshefsky's *The Units of Evolution.*

5. See the papers referred to in note 4 of this chapter for particular versions of these reproductive and genealogical views of species. Within reproductive views, Mayr emphasizes interbreeding, Paterson mate recognition, and Templeton genetic or demographic exchangeability. Within the genealogical views, Cracraft looks to patterns of shared descent, and Wiley to historical fate. For the articulation and debate over genealogical views, see the papers in Quentin D. Wheeler and Rudolf Meier (editors), *Species Concepts and Phylogenetic Theory: A Debate* (New York: Columbia University Press, 2000).

6. I was introduced to the taxonomy of neural crest cells by J. D. Trout, "Attribution, Content, and Method: A Scientific Defense of Commonsense Psychology," Ph.D. thesis, Cornell University, and have drawn in addition on Brian K. Hall and Sven Hörstadius, *The Neural Crest* (Oxford: Oxford University Press, 1988), and Nicole M. Le Douarin, "The Neural Crest," in Gerald Adelman (editor), *Encyclopedia of Neuroscience* (Boston: Birkhauser, 1987), and her *The Neural Crest* (Cambridge: Cambridge University Press, 1982).

7. For the standard taxonomic presentations to which I refer, see the works by Hall and Hörstadius, and Le Douarin cited in note 6 of this chapter.

8. On retinal ganglion cells, extensively studied in cats and frogs, see M. H. Rowe and J. Stone, "Parametric and Feature Extraction Analyses of the Receptive Fields of Visual Neurones: Two Streams of Thought in the Study of a Sensory Pathway," *Brain, Behavior, and Evolution*, 17 (1980), pages 103–122. The quotation from Chalupa is from his "The Nature and Nurture of Retinal Ganglion Cell Development," in Michael Gazzaniga (editor), *The Cognitive Neurosciences* (Cambridge, MA: MIT Press, 1995), page 37.

9. The primary paper of Boyd's on which I shall rely is his "Homeostasis, Species, and Higher Taxa," in Robert A. Wilson (editor), *Species: New Interdisciplinary Essays* (Cambridge, MA: MIT Press, 1999), pages 141–185. See also his, "Realism, Anti-foundationalism, and the Enthusiasm for Natural Kinds," *Philosophical Studies*, 61 (1991), pages 127–148; and "Kinds, Complexity, and Multiple Realization," *Philosophical Studies*, 95 (1999), pages 67–98. For the precursors, see Ludwig Wittgenstein, *Philosophical Investigations* (Oxford: Blackwell, 1953), especially sections 64–71; Hilary Putnam "The 'Analytic' and the

'Synthetic,'" reprinted in his *Mind, Language, and Reality: Philosophical Papers,
Volume 2* (Cambridge: Cambridge University Press, 1975), originally published in 1962; and David Hull, "The Effect of Essentialism on Taxonomy:
Two Thousand Years of Stasis," *British Journal for the Philosophy of Science,* 15
(1965), pages 314–326, and 16 (1965), pages 1–18.

10. On the confirmation of the agreement of morphological and physiological
criteria, see Leo M. Chalupa, "The Nature and Nurture of Retinal Ganglion
Cell Development," pages 40–42.

11. For an alternative way to develop the HPC view of species, and for a corresponding assessment of its broader significance, see Paul E. Griffiths, "Squaring the Circle: Natural Kinds with Historical Essences," in *Species: New Interdisciplinary Essays,* pages 209–228.

12. See Michael Ghiselin, "A Radical Solution to the Species Problem," *Sys-
tematic Zoology,* 23 (1974), pages 536–544, and *Metaphysics and the Origins
of Species* (Albany, NY: SUNY Press, 1997); and David Hull, "Are Species
Really Individuals?," *Systematic Zoology,* 25 (1976), pages 174–191, and "A
Matter of Individuality," *Philosophy of Science,* 45 (1978), pages 335–360.
This paper of Hull's, as well as Ghiselin's, are reprinted in Ereshefsky's *The
Units of Evolution.*

13. For Tyner's original suggestion, see his "The Naming of Neurons: Applications of Taxonomic Theory to the Study of Cellular Populations," *Brain,
Behavior, and Evolution,* 12 (1975), pages 75–96. See also Rowe and Stone,
"Naming of Neurones: Classification and Naming of Cat Retinal Ganglion
Cells," *Brain, Behavior, and Evolution,* 14 (1977), pages 185–216; "The Importance of Knowing One's Own Presuppositions," *Brain, Behavior, and Evolution,* 16 (1979), pages 65–80; "Parametric and Feature Extraction Analyses of the Receptive Fields of Visual Neurones: Two Streams of Thought in
the Study of a Sensory Pathway"; and "The Interpretation of Variation in
the Classification of Nerve Cells," *Brain, Behavior, and Evolution,* 17 (1980),
pages 123–151. See also A. Hughes, "A Rose By Any Other Name: On 'Naming
of Neurones' by Rowe and Stone," *Brain, Behavior, and Evolution,* 16 (1979),
pages 52–64.

14. John Dupré, *The Disorder of Things,* page 57.

15. For my earlier discussion of Dupré's views, see my "Promiscuous Realism,"
British Journal for the Philosophy of Science, 47 (1996), pages 303–316.

16. On the species concept as a union of overlapping concepts, see Philip Kitcher,
"Species," pages 336–337, and for an early discussion of this idea, see David
Hull, "The Effects of Essentialism on Taxonomy: Two Thousand Years of
Stasis."

17. For Sonneborn's view, see especially his monograph-length "Breeding Systems, Reproductive Methods, and Species Problems in Protozoa," in Ernst
Mayr (editor), *The Species Problem* (Washington, DC: American Association
for the Advancement of Science, 1957), pages 155–324.

18. On species and higher taxa, see Marc Ereshefsky, "Species and the Linnaean
Hierarchy," in Robert A. Wilson (editor), *Species: New Interdisciplinary Essays*
(Cambridge, MA: MIT Press, 1999), pages 285–305, and *The Poverty of the
Linnaean Hierarchy* (New York: Cambridge University Press, 2001).

Chapter 6

1. See William Bateson, "The Progress of Genetic Research," delivered as an inaugural address to the Third Conference on Hybridisation and Plant-Breeding in 1906, and reprinted in R. C. Punnett (editor), *The Scientific Papers of William Bateson*, Volume II, (London: Cambridge University Press, 1928), pages 142–151. The quotation is on page 143 of the reprinted version. Johannsen's coinage of "gene" is in his *Elemente der exakten Erblichkeitslehre* (Jena: Gustav Fischer, 1909).

2. See François Jacob, *The Logic of Life* (Princeton, NJ: Princeton University Press, 1973), pages 1–2. The quotation from Dawkins is drawn from *The Selfish Gene* (Oxford: Oxford University Press, 2nd edition, 1989), page 20, originally published in 1976.

3. See Lenny Moss, *What Genes Can't Do* (Cambridge, MA: MIT Press, 2003), especially pages 44–53. The quotation is from page 46.

4. The quotation is from *What Genes Can't Do*, page 47.

5. *What Genes Can't Do*, pages 47–48.

6. The 1994 result was reported by Yoshio Miki et al., "A Strong Candidate for the Breast and Ovarian Cancer Susceptibility Gene BRCA1," *Science*, 266 (1994), pages 66–73.

7. This is a theme in a number of essays in the recent volume edited by Peter Beurton, Raphael Falk, and Hans-Jörg Rheinberger, *The Concept of the Gene in Development and Evolution: Historical and Epistemological Perspectives* (New York: Cambridge University Press, 2000). See especially the chapters by Thomas Fogle, "The Dissolution of Protein Coding Genes in Molecular Biology"; Hans-Jörg Rheinberger, "Gene Concepts: Fragments from the Perspective of Molecular Biology"; and Peter J. Beurton, "A Unified View of the Gene, or How to Overcome Reductionism." See also Petter Portin's useful "The Concept of the Gene: Short History and Present Status," *Quarterly Review of Biology*, 68 (1993), pages 173–223.

8. Keller identifies the "seeming paradox" in Chapter 3 of *The Century of the Gene*, and discusses some of the development of the concept of the gene in Chapter 2 of that book. See Petter Portin, "The Concept of the Gene: Short History and Present Status," for a more detailed history. See also Eva M. Neumann-Held, "Let's Talk about Genes: The Process Molecular Gene Concept and Its Context," in Susan Oyama, Paul E. Griffiths, and Russell D. Gray (editors), *Cycles of Contingency: Developmental Systems and Evolution* (Cambridge, MA: MIT Press, 2001), for a related discussion.

9. See Richard Dawkins, *The Extended Phenotype* (Oxford: Oxford University Press, 1982).

10. *The Extended Phenotype*, page 197.

11. Dawkins makes this "logical point" in *The Extended Phenotype* on pages 198 and 214.

12. Dawkins talks of extended phenotypic effects of replicators at the outset of *The Extended Phenotype*, page 4, while the other pair of quotations are taken from near the end of the second edition of *The Selfish Gene*, from pages 238

and 235, respectively. "The long reach of the gene" is the title of Chapter 13 in the second edition of *The Selfish Gene.*

13. The quotation here is from page 238 of *The Selfish Gene.*

14. On community genetics, see the special editorial feature compiled by Anurag Agrawal in *Ecology*, 84 (2003), pages 543–601. This feature contains lead articles by Claudia Neuhauser et al., and Thomas Whitham et al., together with eight commentaries.

15. The quotation here is from Thomas Whitham et al., "Community and Ecosystem Genetics: A Consequence of the Extended Phenotype," *Ecology*, 84 (2003), pages 559–573, at page 560. See also their Table 1 on the same page for characterizations of some of the central ideas in their version of community genetics.

16. For some discussion of this (and other examples), see "Community and Ecosystem Genetics: A Consequence of the Extended Phenotype," pages 568–570.

17. See Michael J. Wade, "Community Genetics and Species Interactions," *Ecology*, 84 (2003), pages 583–585. The quote is from page 583. See also David Sloan Wilson and William Swenson, "Community Genetics and Community Selection," *Ecology*, 84 (2003), pages 586–588, for the link between community genetics and higher-level selection.

18. The quotation is from François Jacob and Jacques Monod, "Genetic Regulatory Mechanisms in the Synthesis of Proteins," *Journal of Molecular Biology*, 3 (1961), pages 318–356, at page 354.

19. See Evelyn Fox Keller, *Refiguring Life: Metaphors of Twentieth-Century Biology* (New York: Columbia University Press, 1995); *The Century of the Gene* (Cambridge, MA: Harvard University Press, 2000); and "Beyond the Gene but Beneath the Skin," in Susan Oyama, Paul E. Griffiths, and Russell D. Gray (editors), *Cycles of Contingency: Developmental Systems and Evolution* (Cambridge, MA: MIT Press, 2001). The quotation is from page 50 of *The Century of the Gene.*

20. *What Genes Can't Do*, page 2.

21. I discuss encoding views of representation and the contrasting notion of exploitative representation in some detail in Chapter 7 of *Boundaries of the Mind* (New York: Cambridge University Press, 2004).

22. On the polar planimeter and its relevance for thinking about perception and cognition, see Sverker Runeson, "On the Possibility of 'Smart' Perceptual Mechanisms," *Scandinavian Journal of Psychology*, 18 (1977), pages 172–179.

23. Brenner's often-quoted retrospective view of the operon model's view of development is from S. Brenner, W. Dover, I. Herskowitz, and R. Thomas, "Genes and Development: Molecular and Logical Themes," *Genetics*, 126 (1990), pages 479–486, at page 485. For a recent exchange on the use of the informational metaphor in genetics, see John Maynard Smith, "The Concept of Information in Biology," *Philosophy of Science*, 67 (2000), pages 177–194, the accompanying commentaries by Kim Sterelny, Peter Godfrey-Smith, and Sahotra Sarkar, and the subsequent discussion of Paul E. Griffiths, "Genetic Information: A Metaphor in Search of a Theory," *Philosophy of Science*, 68 (2001), pages 394–412.

Chapter 7

1. See Erwin Schrödinger, *What is Life?* (Cambridge: Cambridge University Press, 1944). The quotation is from page 19.
2. I have taken this statement of the sequence hypothesis from Lily E. Kay's *Who Wrote the Book of Life? A History of the Genetic Code* (Palo Alto, CA: Stanford University Press, 2000), page 174. It also appears on page 52 of Evelyn Fox Keller's *The Century of the Gene* (Cambridge, MA: Harvard University Press, 2000). For Crick's original statement, see his "On Protein Synthesis," in *Symposium of the Society for Experimental Biology*, 12 (1958), pages 138–163, at pages 152–153.
3. The figure of 1.5% is taken from David Baltimore, "Our Genome Unveiled," *Nature*, 409 (2001), pages 814–816. The quotation from Keller is taken from *The Century of the Gene*, pages 58–59.
4. For an insider version of the story of the discovery and significance of the homeobox, see Walter J. Gehring, *Master Control Genes in Development and Evolution: The Homeobox Story* (New Haven, CT: Yale University Press, 1998). I have drawn on Chapters 2 and 3 especially in this paragraph. The phrase "master control genes," as well as the quotation that follows, are from page xi of the preface.
5. See Denis Walsh, "Alternative Individualism," *Philosophy of Science*, 66 (1999), pages 628–648. The quotation is from page 640.
6. For the details provided in this closer look, I have relied largely on Chapter 16 of Scott Gilbert's, *Developmental Biology* (Cambridge, MA: Sinauer Associates, 5th edition, 1997). See also Gehring's *Master Control Genes in Development and Evolution: The Homeobox Story*, for further details.
7. Aficionados of the debate between individualism and externalism in the philosophy of mind may know that I have been exercised by the claim that scientific taxonomy is "taxonomy by causal powers" in other contexts. For a classic statement of this claim, see Jerry A. Fodor, *Psychosemantics: The Problem of Meaning in the Philosophy of Mind* (Cambridge, MA: MIT Press, 1987), Chapter 2. For a response, see Robert A. Wilson, *Cartesian Psychology and Physical Minds: Individualism and the Sciences of the Mind* (New York: Cambridge University Press), Chapter 2. These chapters are reprinted together in William G. Lycan (editor), *Mind and Cognition: A Reader* (New York: Blackwell, 2nd edition, 1998).
8. See Susan Oyama, *The Ontogeny of Information: Developmental Systems and Evolution* (Durham, NC: Duke University Press, 2nd edition, 2000), as well as a collection of her essays, *Evolution's Eye: A Systems View of the Biology-Culture Divide* (Durham, NC: Duke University Press, 2000). The quotation from Kitcher is from his "Battling the Undead: How (and How Not) to Resist Genetic Determinism," in R. S. Singh, C. B. Krimbas, D. B. Paul, and J. Beatty (editors), *Thinking About Evolution: Historical, Philosophical, and Political Perspectives* (New York: Cambridge University Press, 2001), page 408. This essay is also reprinted in Kitcher's recent collection, *In Mendel's Mirror: Philosophical Reflections on Biology* (New York: Oxford University Press, 2003). For a recent, direct response to Kitcher, see Paul E. Griffiths, "The Fearless Vampire Conservator: Philip Kitcher, Genetic Determinism and the Informational Gene," in

C. Rehmann-Sutter and E. M. Neumann-Held (editors), *Genes in Development* (Durham, NC, Duke University Press, in press).

9. Much of the emerging field of evolutionary developmental biology, or "evo-devo" takes these critical points for granted and explores various developmental systems. For a brief discussion of the relationship between evo-devo and DST, see Jason Scott Robert, Brian K. Hall, and Wendy M. Olson, "Bridging the Gap Between Developmental Systems Theory and Evolutionary Developmental Biology," *BioEssays*, 23 (2001), pages 954–962. For a popular, general treatment of some of these issues, see David S. Moore, *The Dependent Gene: The Fallacy of 'Nature' vs. 'Nurture'* (New York: Henry Holt, 2001).

10. Evelyn Fox Keller's views here are articulated in her "Beyond the Gene but Beneath the Skin," in Susan Oyama, Paul E. Griffiths, and Russell D. Gray (editors), *Cycles of Contingency: Developmental Systems and Evolution* (Cambridge, MA: MIT Press, 2001), pages 299–312. The work of Jablonka and Lamb that I have in mind is their *Epigenetic Inheritance and Evolution: the Lamarckian Dimension* (New York: Oxford University Press, 1995); see also their target review article "Epigenetic Inheritance in Evolution," *Journal of Evolutionary Biology*, 11 (1998), pages 159–183.

11. On computationalism and individualism, see Robert A. Wilson, *Boundaries of the Mind* (New York: Cambridge University Press, 2004), especially Chapters 7–8. "Wide computationalism" was coined and first presented as a challenge to the inference from computationalism to individualism about the mind in my "Wide Computationalism," *Mind*, 103 (1994), pages 351–372.

12. Expressions of DST that make this sort of move include Paul E. Griffiths and Russell D. Gray, "Developmental Systems and Evolutionary Explanation," *Journal of Philosophy*, 91 (1994), pages 277–304, reprinted in David Hull and Michael Ruse (editors), *The Philosophy of Biology* (New York: Oxford University Press, 1998); Griffiths and Gray, "Darwinism and Developmental Systems," in *Cycles of Contingency*, pages 195–218; Russell Gray, "Selfish Genes or Developmental Systems? Evolution without Interactors and Replicators?," in *Thinking about Evolution*, pages 184–207; and Tim Ingold, "From Complementarity to Obviation: On Dissolving the Boundaries Between Social and Biological Anthropology, Archaeology, and Psychology," in *Cycles of Contingency*, pages 255–279.

13. These quotations are taken from their "Developmental Systems and Evolutionary Explanation," pages 130 and 131 of the version reprinted in David Hull and Michael Ruse (editors), *The Philosophy of Biology* (New York: Oxford University Press, 1998).

14. Paul Griffiths and Russell Gray discuss both the methylation system and that of *Buchnera* bacteria in their "Darwinism and Developmental Systems," pages 197–198. The former of these receives detailed treatment in the work of Jablonka and Lamb already mentioned (see note 10, this chapter).

15. On birdsong, see Peter Marler, "Song Learning: Innate Species Differences in the Learning Process," in Peter Marler and Herbert S. Terrace (editors), *The Biology of Learning* (Berlin: Springer, 1984); and "Differences in Behavioural Development in Closely Related Species: Birdsong," in Patrick Bateson (editor), *The Development and Integration of Behaviour* (Cambridge: Cambridge

University Press, 1991). On social play, see Mark Bekoff, "Play Signals as Punctuation: The Structure of Social Play in Canids," *Behaviour*, 132 (1995), pages 419–429; and "Playing with Play: What Can We Learn About Evolution and Cognition," in Denise Dellarosa Cummins and Colin Allen (editors), *The Evolution of Mind* (New York: Oxford University Press, 1998).

16. The notion of the ambient optical array is central to the ecological approach to perception articulated and defended by James J. Gibson. See, for example, his *The Ecological Approach to Visual Perception* (Boston, MA: Houghton-Mifflin, 1979). On external storage systems, see Merlin Donald, *The Origins of the Modern Mind* (Cambridge, MA: Harvard University Press, 1991). And on distributed cognition, see Edwin Hutchins, *Cognition in the Wild* (Cambridge, MA: MIT Press, 1995).

17. The termite mound example is taken from Chapter 11 of J. Scott Turner, *The Extended Organism: The Physiology of Animal-Built Structures* (Cambridge, MA: Harvard University Press, 2000). See also my earlier discussion of his examples of living coral reefs in Chapter 3 and of the idea of what I call corporate organisms in Chapter 4.

18. See Turner, *The Extended Organism*; Kevin N. Laland, F. John Odling-Smee, and Marcus W. Feldman, "Niche Construction, Biological Evolution, and Cultural Change," *Behavioral and Brain Sciences*, 23 (2000), pages 131–175; F. John Odling-Smee, Kevin N. Laland, and Marcus W. Feldman, *Niche Construction: The Neglected Process in Evolution* (Princeton, NJ: Princeton University Press, 2003); and Jablonka and Lamb, *Epigenetic Inheritance and Evolution*.

19. See Richard Lewontin, "Gene, Organism, and Environment," in D. S. Bendall (editor), *Evolution from Molecules to Men* (Cambridge: Cambridge University Press, 1983), pages 273–285. The quotation is from page 280. See also Lewontin's recent *The Triple Helix: Gene, Organism, and Environment* (Cambridge, MA: Harvard University Press, 2000).

20. *Niche Construction*, page 41.

21. *The Extended Organism*, Chapter 10. The quotation is from page 178.

22. The quotation is from *Niche Construction*, page 178. For their discussion of ecological inheritance, see especially pages 12–16.

23. On epigenetic inheritance, see Jablonka and Lamb, *Epigenetic Inheritance and Evolution* and "Epigenetic Inheritance in Evolution." On behavioral inheritance, see E. Avital and E. Jablonka, *Animal Traditions: Behavioural Inheritance in Evolution* (Cambridge: Cambridge University Press, 2000). The more recent work of Jablonka's I refer to is her "The Systems of Inheritance," in *Cycles of Contingency*, pages 99–116. See also Eva Jablonka and Marion Lamb, *Evolution in Four Dimensions* (Cambridge, MA: MIT Press, 2005).

24. For the quotation, see Oyama, Griffiths, and Gray, "Introduction: What is Developmental Systems Theory?," in *Cycles of Contingency*, pages 1–11, at page 6.

Chapter 8

1. On the inroads of evolutionary approaches in psychology and anthropology, see Jerome Barkow, Leda Cosmides, and John Tooby (editors), *The Adapted*

Mind: Evolutionary Psychology and Generation of Culture (New York: Oxford University Press, 1992); in medicine, see Randolph M. Nesse and George C. Williams, *Why We Get Sick: The New Science of Darwinian Medicine* (New York: Times Books, 1994); in epidemiology, see Paul Ewald, *Plague Time: The New Germ Theory of Disease* (New York: Anchor Books, 2002).

2. Stephen Jay Gould, *The Structure of Evolutionary Theory* (Cambridge, MA: Harvard University Press, 2002), pages 126 and 127.

3. For the quotation, see *The Structure of Evolutionary Theory*, page 634, as well as Gould's "The Evolutionary Definition of Selective Agency, Validation of the Theory of Hierarchical Selection, and Fallacy of the Selfish Gene," in Rama S. Singh, Costas B. Krimbas, Diane B. Paul, and John Beatty (editors), *Thinking About Evolution: Historical, Philosophical, and Political Perspectives* (New York: Cambridge University Press, 2001), page 225.

4. See Richard Lewontin, "The Units of Selection," *Annual Review of Ecology and Systematics*, 1 (1970), pages 1–18.

5. For the Futuyma quote, see his textbook *Evolutionary Biology*, 2nd edition (Sunderland, MA: Sinauer, 1986), page 554.

6. See David Hull, "Individuality and Selection," *Annual Review of Ecology and Systematics*, 11 (1980), pages 311–332, at page 318.

7. For a recent, otherwise admirable discussion of replicators that seems to me too quick to dismiss organismal replication, see Peter Godfrey-Smith, "The Replicator in Retrospect," *Biology and Philosophy*, 15 (2000), pages 403–423.

8. See Elisabeth Lloyd, "Units and Levels of Selection: An Anatomy of the Units of Selection Debates," in *Thinking About Evolution: Historical, Philosophical, and Political Perspectives* (New York: Cambridge, 2001), pages 267–291.

9. Richard Dawkins, *The Extended Phenotype* (Oxford: Oxford University Press, 1982), page 4.

10. The quotations from Williams is on page 251 of *Adaptation and Natural Selection*, and that from Dawkins on page viii of the second edition of *The Selfish Gene*. For Hamilton's classic papers, see his "The Evolution of Altruistic Behavior," *American Naturalist*, 97 (1963), pages 354–356; "The Genetical Evolution of Social Behaviour I," *Journal of Theoretical Biology*, 7 (1964), pages 1–16; and "The Genetical Evolution of Social Behaviour II," *Journal of Theoretical Biology*, 7 (1964), pages 17–52. These three papers are reprinted, together with extended introductory comments by Hamilton, in the first volume of his collected papers, *Narrow Roads of Gene Land, Volume 1: Evolution of Social Behavior* (New York: Freeman, 1996).

11. Charles Darwin, *The Descent of Man, and Selection in Relation to Sex* (Princeton, NJ: Princeton University Press, 1981), originally published in 1871. The quotation is on page 166.

12. The quotation is from Alfred Emerson, "Ecology, Evolution and Society," *American Naturalist*, 77 (1943), pages 97–118, at page 118. For Emerson's more extended views of the superorganism, see his "Social Coordination and the Superorganism," *American Midland Naturalist*, 21 (1939), pages 182–209; "Basic Comparisons of Human and Insect Societies," *Biological Symposia*, 8 (1942), pages 163–176; and "The Biological Basis of Social Cooperation," *Illinois Academy of Science Transactions*, 29 (1946), pages 9–18.

13. The quotation is drawn by Hamilton from V. B. Wigglesworth, *The Life of Insects* (London: Wiedenfeld and Nicolson, 1964), and is given on *Narrow Roads of Gene Land*, page 22.

14. *Adaptation and Natural Selection*, pages 4–5.

15. The quote is from page 92 of *Adaptation and Natural Selection*. Williams's later, somewhat more liberal views about group selection are given in *Natural Selection: Domains, Levels, and Challenges* (New York: Oxford, 1992).

16. For Hamilton's original view of sex ratios, see his "Extraordinary Sex Ratios," *Science*, 156 (1967), pages 477–488, reprinted in *Narrow Roads of Gene Land*.

17. For the joint work of most influence, see their "Reviving the Superorganism," *Journal of Theoretical Biology*, 136 (1989), pages 337–356; "A Critical Review of Philosophical Work on the Units of Selection Problem," *Philosophy of Science*, 61 (1994), pages 534–555; "Reintroducing Group Selection to the Human Behavioral Sciences," *Behavioral and Brain Sciences*, 17 (1994), pages 585–654; and *Unto Others: The Evolution and Psychology of Unselfish Behavior* (Cambridge, MA: Harvard University Press, 1998). For one place in which they express agreement with the Matching Principle, see "Reintroducing Group Selection to the Human Behavioral Sciences," page 597. Wilson's early work is represented by his "A Theory of Group Selection," *Proceedings of the National Academy of Sciences USA*, 72 (1975), pages 143–146; "Structured Demes and the Evolution of Group Advantageous Traits," *American Naturalist*, 111 (1977), pages 157–185; *The Natural Selection of Populations and Communities* (Menlo Park, CA: Benjamin Cummins, 1980); and "The Group Selection Controversy: History and Current Status," *Annual Review of Ecology and Systematics*, 14 (1983), pages 159–187. For his more recent work, see his "Altruism and Organism: Disentangling the Themes of Multilevel Selection Theory," *American Naturalist*, 150 supp. (1997), pages S122–S134; "Incorporating Group Selection into the Adaptationist Program: A Case Study Involving Human Decision Making," in J. Simpson and D. Kendrick (editors), *Evolutionary Social Psychology* (Hillsdale, NJ: Erlbaum, 1997); "A Critique of R. D. Alexander's Views on Group Selection," *Biology and Philosophy*, 14 (1999), pages 431–449; and *Darwin's Cathedral: Evolution, Religion, and the Nature of Society* (Chicago: University of Chicago Press, 2002).

18. The quote from E. O. Wilson is from page 579 of *Sociobiology: The New Synthesis* (Cambridge, MA: Harvard University Press, 2000 edition, first published in 1975). See also Alexander Rosenberg, "Altruism: Theoretical Contexts," in Evelyn Fox Keller and Elisabeth A. Lloyd (editors), *Keywords in Evolutionary Biology* (Cambridge, MA: Harvard University Press, 1992).

19. For the original discussion of reciprocal altruism, see Robert Trivers, "The Evolution of Reciprocal Altruism," *Quarterly Review of Biology*, 46 (1971), pages 35–57.

20. The first and last quotations in this paragraph are taken from pages 257 and 260, respectively, of Darwin's *Origin*; the middle two quotations are on page 258.

21. I have in mind especially the work of Elliott Sober and David Sloan Wilson on multilevel selection. See especially their *Unto Others*, and David Sloan Wilson's "Altruism and Organism: Disentangling the Themes of Multilevel

Selection Theory" and, "Incorporating Group Selection into the Adaptationist Program: A Case Study Involving Human Decision Making."

22. The general parallel between the conditions for individual and group selection are clear in Richard Lewontin's "The Units of Selection" and David Sloan Wilson's "The Group Selection Controversy: History and Current Status," see especially page 170. I shall discuss Michael Wade's work in Chapter 10, but for two of the initial papers, see his "An Experimental Study of Group Selection," *Evolution*, 31 (1977), pages 134–153; and "A Critical Review of the Models of Group Selection," *Quarterly Review of Biology*, 53 (1978), pages 101–114.

23. The quotation from Dawkins here is from *The Selfish Gene*, 2nd edition, page 10.

24. The characterization of a trait group is given on page 143 of Wilson's "A Theory of Group Selection," as are the examples I mention. Talk of the "sphere of influence" of a trait can be found on pages 20–24 of his *The Natural Selection of Populations and Communities*, and Chapters 5–6 of that book explore evolution in multispecies communities.

25. The slightly different characterization of a trait group is given on page 92 of *Unto Others*. The *Wolbachia* example is discussed by Michael J. Wade, "Infectious Speciation," *Nature*, 409 (2001), pages 675–677; and that of mites and carrion beetles by David Sloan Wilson, "Holism and Reductionism in Evolutionary Ecology," *Oikos*, 53 (1988), pages 269–273.

26. Kim Sterelny, "The Return of the Group," *Philosophy of Science*, 63 (1996), pages 562–584; the discussion is on pages 565–567.

27. Wilson first discussed the butterfly example in his "A Theory of Group Selection," and Sober and Wilson return to discuss it on page 94 of *Unto Others*.

28. The quotations are from pages 605 and 598, respectively, of "Reintroducing Group Selection to the Human Behavioral Sciences," and the idea that groups become organism-like through group selection is endorsed on page 599.

29. On the idea of human social groups as superorganisms, see not only "Reintroducing Group Selection . . . ," pages 602 and forward, but also Wilson's "Altruism and Organism: Disentangling the Themes of Multilevel Selection Theory" and *Darwin's Cathedral*; and D. S. Wilson, C. Wilczynski, A. Wells, and L. Weiser, "Gossip and Other Aspects of Language as Group-Level Adaptations," in Celia Heyes and Ludwig Huber (editors), *Evolution and Cognition* (Cambridge, MA: MIT Press, 2000).

30. The quotation is from "Reintroducing Group Selection to the Human Behavioral Sciences," page 592. Another implication of this use of the framework is that organs are a kind of organism, a claim developed in different ways by Leo Buss, *The Evolution of Individuality* (Princeton, NJ: Princeton University Press, 1987); and by Richard E. Michod *Darwinian Dynamics: Evolutionary Transitions in Fitness and Individuality* (Princeton, NJ: Princeton University Press, 1999). See Chapters 3 and 4 for why I think this is a mistaken view.

31. See Steven Stanley, *Macroevolution: Pattern and Process* (San Francisco: W. H. Freeman, 1979); and Steven Jay Gould, *The Structure of Evolutionary Theory*, for example.

32. The quotation comes from Kim Sterelny, "Last Will and Testament: Stephen Jay Gould's *The Structure of Evolutionary Theory*," *Philosophy of Science*, 70 (2003), pages 255–263, at page 260. It expresses the stronger constraint introduced by Vrba in her "What is Species Selection?" *Systematic Zoology*, 33 (1984), pages 318–328; and "Levels of Selection and Sorting with Special Reference to the Species Level," in Paul H. Harvey and Linda Partridge (editors), *Oxford Surveys in Evolutionary Biology, Volume 6* (Oxford: Oxford University Press, 1989), pages 111–168. For paleontologists who operate with weaker notions of species and clade selection, see for example Steven Stanley, *The New Evolutionary Timetable: Fossils, Genes, and the Origin of Species* (New York: Basic Books, 1981), pages 185–186.

33. Accounts of evolution of planktotrophic mollusks that appeal to higher level selection are offered by David Jablonski, "Larval Ecology and Macroevolution in Marine Invertebrates," *Bulletin of Marine Sciences* 39 (1986), pages 565–587; and "Heritability at the Species Level: Analysis of Geographic Ranges of Cretaceous Mollusks," *Science*, 238 (1987), pages 360–363; of body size in mammals by J. H. Brown and B. A. Maurer, "Evolution of Species Assemblages," *American Naturalist*, 130 (1987), pages 1–17; and "Macroecology": The Division of Food and Space Among Species on Continents," *Science*, 243 (1989), pages 1145–1150; and of flowering plants by Steven M. Stanley, *The New Evolutionary Timetable*, pages 90–91.

34. I draw on Stanley's Chapter 7 more generally for the following discussion. See also Stephen Jay Gould, *The Structure of Evolutionary Theory*, pages 717–719 for his three-page table, and pages 703–744 for his more general discussion of "the grand analogy" between organismal and species selection.

35. See Todd Grantham, "Hierarchical Approaches to Macroevolution: Recent Work on Species Selection and the 'Effect Hypothesis,'" *Annual Review of Ecology and Systematics*, 26 (1995), pages 301–321.

36. Apart from the works cited in note 32, see also Elisabeth Vrba and Stephen Jay Gould, "The Hierarchical Expansion of Sorting and Selection: Sorting and Selection Cannot Be Equated," *Paleobiology*, 12 (1986), pages 217–228.

37. For Sober's initial distinction, see his *The Nature of Selection* (Cambridge, MA: MIT Press, 1984), pages 97–102. For Vrba's antipathy to what Sober does with it, see especially her "Levels of Selection and Sorting with Special Reference to the Species Level," pages 127–129.

Chapter 9

1. For two ways of dividing up the traditions of group selection, see David Sloan Wilson, "The Group Selection Controversy: History and Current Status," *Annual Review of Ecology and Systematics*, 14 (1983), pages 159–187; and Charles Goodnight and Lori Stevens, "Experimental Studies of Group Selection: What Do They Tell Us About Group Selection in Nature?," *American Naturalist*, 150 (supp.) (1997), pages S59–S79.

2. For Lewontin's introduction, see his "The Units of Selection," *Annual Review of Ecology and Systematics*, 1 (1970), pages 1–18. Sober and Wilson discuss this case in several places, but I will focus on their most sustained, recent

discussion in their *Unto Others: The Evolution and Psychology of Unselfish Behavior* (Cambridge, MA: Harvard University Press, 1998), pages 43–50.

3. *Unto Others*, pages 35–50.

4. The quotation is from page 46 of *Unto Others*.

5. For an example of this claim about genic selection, see Kim Sterelny, "The Return of the Group," *Philosophy of Science*, 63 (1996), pages 562–584.

6. Now classic statements of pluralism include those of Lee Alan Dugatkin and Hudson K. Reeve, "Behavioral Ecology and Levels of Selection: Dissolving the Group Selection Controversy," in Peter J. B. Slater et al. (editors), *Advances in the Study of Behavior*, Vol. 23 (New York: Academic Press, 1994); and Kim Sterelny, "The Return of the Group." For further references, see Chapter 10.

7. The quotation is from page 15 of "The Units of Selection."

8. The quotations in this paragraph are from David Sloan Wilson, "A Theory of Group Selection," *Proceedings of the National Academy of Sciences*, 72 (1975), pages 143–146 at page 143; and *Unto Others*, page 92. See also D. S. Wilson, "Levels of Selection: An Alternative to Individualism in the Human Sciences," *Social Networks*, 11 (1989), pages 257–272, reprinted in Elliott Sober (editor), *Conceptual Issues in the Philosophy of Biology* (Cambridge, MA: MIT Press, 2nd edition, 1993), pages 143–154, especially page 148; and Elliott Sober, *Philosophy of Biology* (Boulder, CO: Westview, 1993), page 109.

9. See Douglas Futuyma, *Evolutionary Biology* (Sunderland, MA: Sinauer, 1979), page 455; and Richard Alexander and Gerald Borgia, "Group Selection, Altruism, and the Levels of Organization of Life," *Annual Review of Ecological Systematics*, 9 (1978), pages 449–474, especially pages 451–453. Later editions of Futuyma's text, such as the second (1986) and the third (1998) introduce the group selection explanation as a possibility, though not one that Futuyma himself seems to take a stand on there.

10. For Sober and Wilson's criticisms, see *Unto Others*, pages 31–34 and 46–50. For Sober's original criticisms, see his *The Nature of Selection* (Cambridge, MA: MIT Press, 1984), pages 327–335. And for the claim that the averaging approach vitiates the framework for debate between individual and group selectionists, see David Sloan Wilson, "Levels of Selection: An Alternative to Individualism in the Human Sciences."

11. See David Lack, *The Natural Regulation of Animal Numbers* (New York: Oxford University Press, 1954).

12. See Frank Fenner and F. N. Ratcliffe, *Myxomatosis* (Cambridge: Cambridge University Press, 1965); and Frank Fenner and Bernadino Fantini, *Biological Control of Vertebrate Pests: The History of Myxomatosis – an Experiment in Evolution* (New York: CABI Publishing, 1999).

13. See *Myxomatosis*, pages 284–286.

14. *Myxomatosis*, page 221 for the quotation, and Table 30, page 163 for supporting data.

15. See I. Parer, W. R. Sobey, D. Conolly, and R. Morton, "Sire Transmission of Acquired Resistance to Myxomatosis," *Australian Journal of Zoology*, 43 (1995), pages 459–465. The quotation is on page 462. This work builds on that of Sobey and Conolly, "Myxomatosis: Non-genetic Aspects of Resistance to Myxomatosis in Rabbits, *Oryctolagus cuniculus*," *Australian Wildlife Research*, 13

(1986), pages 177–187 on domestic rabbits; and C. K. Williams and R. J. Moore, "Inheritance of Acquired Immunity to Myxomatosis," *Australian Journal of Zoology*, 39 (1991), pages 307–311 on wild rabbits.

16. Ian Parer, "Relationship Between Survival Rate and Survival Time of Rabbits, *Oryctolagus cuniculus* (L.), Challenged with Myxoma Virus," *Australian Journal of Zoology*, 43 (1995), pages 303–311. The quotation is drawn from page 308. See also the very brief discussion by Fenner and Fantini in their *Biological Control of Vertebrate Pests*, pages 193 and 297–298.

17. For Fenner and Ratcliffe's identification, see *Myxomatosis*, pages 222–223. See also *Biological Control of Vertebrate Pests*, pages 200–201 and 300–301 on the idea of latent infections and the problem of overwintering.

18. For the numbers on these inoculation campaigns, see *Myxomatosis*, page 292, Table 50.

19. The quotation is from *Myxomatosis*, page 222, and their later discussion of attenuated strains on page 341.

20. The general treatments I have in mind are those by Paul Schmid-Hempel, *Parasites in Social Insects* (Princeton, NJ: Princeton University Press, 1998), at page 251; and by Paul Ewald, *The Evolution of Infectious Disease* (New York: Oxford University Press, 1994), Chapter 3; and more recently his *Plague Time: The New Germ Theory of Disease* (New York: Anchor Books, 2002), pages 27–31. Andrew Levine, *Viruses* (New York: Scientific American Library, 1992), pages 209–210, suggests a view closer to that which I am tentatively ascribing to Fenner and his coauthors.

21. For Wade's experiments, see for example his "An Experimental Study of Group Selection," *Evolution*, 31 (1977), pages 134–153; "A Critical Review of the Models of Group Selection," *Quarterly Review of Biology*, 53 (1978), pages 101–114; and "Group Selection: Migration and the Differentiation of Small Populations," *Evolution*, 36 (1982), pages 944–961.

22. For reports from the Canberra and Lake Urana studies, see Fenner and Fantini's *Biological Control of Vertebrate Pests*, pages 183–185.

23. Fenner and Ratcliffe discuss what I am calling an *E.coli*-like strain in myxoma in *Myxomatosis*, pages 346–347.

24. See Kim Sterelny and Philip Kitcher, "The Return of the Gene," *Journal of Philosophy*, 85 (1988), pages 339–361; and Elliott Sober and Richard Lewontin, "Artifact, Cause, and Genic Selection," *Philosophy of Science*, 49 (1982), pages 157–180.

25. For examples, see Lewontin's "The Units of Selection," page 15; and Sober and Wilson's *Unto Others*, page 45. See also Samir Okasha, "Genetic Relatedness and the Evolution of Altruism," *Philosophy of Science*, 69 (2002), pages 138–149 for a recent, effective example of genic viewpoint talk.

26. For Fenner and Ratcliffe's evidence, see *Myxomatosis*, pages 233–235.

Chapter 10

1. Although Sober and Wilson present their multilevel view of natural selection in a variety of publications, I shall concentrate on the discussions to be found

in their "Reintroducing Group Selection to the Human Behavioral Sciences," *Behavioral and Brain Sciences*, 17 (1994), pages 585–654; and *Unto Others: The Evolution and Psychology of Unselfish Behavior* (Cambridge, MA: Harvard University Press, 1998).

2. Something like these two forms of pluralism are recognized by Sober and Wilson in *Unto Others*, page 331, and also by Paul E. Griffiths, "Review of Sober and Wilson 1998," *Mind*, 111 (2002), pages 178–182.

3. Both the longer and the shorter quotations are from *Unto Others*, page 98.

4. The expressions of model pluralism I have in mind are those of Kim Sterelny, "The Return of the Group," *Philosophy of Science*, 63 (1996), pages 562–584; of Lee Alan Dugatkin and Hudson K. Reeve, "Behavioral Ecology and Levels of Selection: Dissolving the Group Selection Controversy," in Peter J. B. Slater et al. (editors), *Advances in the Study of Behavior*, Vol. 23 (New York: Academic Press, 1994), pages 101–133; and most recently of Benjamin Kerr and Peter Godfrey-Smith, "Individualist and Multi-Level Perspectives on Selection in Structured Populations," *Biology and Philosophy*, 17 (2002), pages 477–517. The Kerr and Godfrey-Smith paper appears with commentaries by Lee Alan Dugatkin, John Maynard Smith, and Elliott Sober and David Sloan Wilson, together with the author's reply and a complementary paper. Dugatkin and Reeve's characterization of broad-sense individualism is given on page 107 of their paper.

5. For those working on social insects and eusociality from this perspective, see Andrew F. G. Bourke and Nigel R. Franks, *Social Evolution in Ants* (Princeton, NJ: Princeton University Press, 1995), and Ross H. Crozier and Pekka Pamilo, *Evolution of Social Insect Colonies* (New York: Oxford University Press, 1996). Those working on the evolution of multicellular and social life include Steve A. Frank, *Foundations of Social Evolution* (Princeton, NJ: Princeton University Press, 1998), and Richard Michod, *Darwinian Dynamics: Evolutionary Transitions in Fitness and Individuality* (Princeton, NJ: Princeton University Press, 1999). Apart from the work of Dugatkin and Reeve referred to in note 4, see also Alan Grafen, "Natural Selection, Kin Selection, and Group Selection," in J. Krebs and N. Davies (editors), *Behavioural Ecology: An Evolutionary Approach* (London: Blackwell, 1984), pages 63–84; as well as the short discussions of Hudson K. Reeve, "Multi-Level Selection and Human Cooperation," *Evolution and Human Behavior*, 21 (2000), pages 65–72; and Lee Alan Dugatkin, "Will Peace Follow?," *Biology and Philosophy*, 17 (2002), pages 519–522. The quotation from Bourke and Franks is from page 67 of their book.

6. The idea that groups form part of an individual's (or individual gene's) environment can be found both in Kim Sterelny and Philip Kitcher, "The Return of the Gene," *Journal of Philosophy*, 85 (1988), pages 339–361, and in Richard Dawkins, *The Selfish Gene*, 2nd edition (Oxford: Oxford University Press, 1989), at page 258. Sterelny suggests the range of cases I mention in "The Return of the Group," pages 570–573. The last of these, that of trematode altruism, is taken from Richard Dawkins, *The Extended Phenotype* (Oxford: Oxford University Press, 1982).

7. Dawkins appeals to the Necker cube analogy in the first chapter of *The Extended Phenotype*, and again in the preface to the second edition of *The Selfish Gene*.

For Kitcher's Neckerphilia, see for example his recent "Evolutionary Theory and the Social Uses of Biology," *Biology and Philosophy*, 19 (2004), pages 1–15, and for that of Kerr and Godfrey-Smith, "Individualist and Multi-Level Perspectives on Selection in Structured Populations."

8. The quotations are from Dugatkin and Reeve's "Behavioral Ecology and Levels of Selection: Dissolving the Group Selection Controversy," page 108, and Sterelny's "The Return of the Group," page 572.

9. "Multi-Level Selection and Human Cooperation," pages 65–66.

10. See "The Return of the Group," page 583 for both the quotation and for the rejection of trait groups as vehicles.

11. All of the quotes in this paragraph are from "Behavioral Ecology and Levels of Selection," pages 108–109.

12. See "Individualist and Multi-Level Perspectives on Selection in Structured Populations," page 481 for the quotation and pages 482–484 for the supporting discussion of dynamic sufficiency.

13. The quotation is from "Behavioral Ecology and Levels of Selection," page 109, italics in the original. See also Bourke and Franks, *Social Evolution in Ants*, pages 45–49, especially Box 2.1.

14. On the use of cellular automata in modeling the immune system, see A. T. Bernardes and R. M. Zorzenon Dos Santos, "Immune Network at the Edge of Chaos," *Journal of Theoretical Biology*, 186 (1997), pages 173–187.

15. Dugatkin and Reeve provide their mapping on page 120, Table 1. The quotation from Kerr and Godfrey-Smith can be found on page 484 of their "Individualist and Multi-Level Perspectives on Selection in Structured Pupolations."

16. On the rarity of group selection, see not only Sterelny on trait groups, "The Return of the Group," but also John Maynard Smith, "Commentary on Kerr and Godfrey-Smith," *Biology and Philosophy*, 17 (2002), pages 523–527, and the classic work of George C. Williams, *Adaptation and Natural Selection: A Critique of Some Current Evolutionary Thought* (Princeton, NJ: Princeton University Press, 1966).

17. On pluralism and realism in this context, see Elliott Sober and David Sloan Wilson, "Perspectives and Parameterizations: Commentary on Benjamin Kerr and Peter Godfrey-Smith's 'Individualist and Multi-Level Perspectives on Selection in Structured Populations,' " *Biology and Philosophy*, 17 (2002), pages 529–537.

18. For Wade's classic work, see his "An Experimental Study of Group Selection," *Evolution* 31 (1977), pages 134–153; "A Critical Review of the Models of Group Selection," *Quarterly Review of Biology*, 53 (1978), pages 101–114; "An Experimental Study of Kin Selection," *Evolution*, 34 (1980), pages 844–855; and "Group Selection: Migration and the Differentiation of Small Populations," *Evolution*, 36 (1982), pages 944–961. For the related studies on emigration rate, see D. M. Craig, "Group Selection Versus Individual Selection: An Experimental Analysis," *Evolution*, 36 (1982), pages 271–282, and on leaf area, Charles Goodnight, "The Influence of Environmental Variation on Group and Individual Selection in a Cress," *Evolution*, 39 (1985), pages 545–558.

19. Charles Goodnight and Lori Stevens, "Experimental Studies of Group Selection: What Do They Tell Us About Group Selection in Nature?," *American Naturalist*, 150 (1997), pages S59–S79. The quotation is on page S67.

20. The quotation is from "The Return of the Gene," page 571.

21. On indirect genetic effects, see Anurag F. Agrawal, Edmund D. Brodie III, and Michael J. Wade, "On Indirect Genetic Effects in Structured Populations," *American Naturalist*, 158 (2001), pages 308–323; and Jason B. Wolf, Edmund D. Brodie III, and Michael J. Wade, "Evolution When the Environment Contains Genes," in Thomas J. DeWitt and Samuel M. Scheiner (editors), *Phenotypic Plasticity: Functional and Conceptual Approaches* (New York: Oxford University Press, 2004).

22. For a recognition of something like entwinement, and for techniques to distinguish apparently entwined properties, see I. Lorraine Heisler and John Damuth, "A Method for Analyzing Selection in Hierarchically Structured Populations," *American Naturalist*, 130 (1987), pages 582–602; and John Damuth and I. Lorraine Heisler, "Alternative Formulations of Multilevel Selection," *Biology and Philosophy*, 3 (1988), pages 407–430.

23. For the distinction of levels in terms of scale in physics, see Max Dresden, "The Klopsteg Memorial Lecture: Fundamentality and Numerical Scales – Diversity and the Structure of Physics," *American Journal of Physics*, 66 (1998), pages 468–482.

24. Black's work on metaphor is represented in the essays in his *Models and Metaphors* (Ithaca, NY: Cornell University Press, 1962). See also Mary Hesse, *Models and Analogies in Science* (South Bend, IN: University of Notre Dame Press, 1966), and Richard N. Boyd, "Metaphor and Theory Change: What is 'Metaphor' a Metaphor For?," in Andrew Ortony (editor), *Metaphor and Thought*, 2nd edition (Cambridge: Cambridge University Press, 1993), originally published in 1979.

25. The quotation is from Hudson K. Reeve and Laurent Keller, "Levels of Selection: Burying the Units-of-Selection Debate and Unearthing the Crucial New Issues," in Laurent Keller (editor), *Levels of Selection in Evolution* (Princeton, NJ: Princeton University Press, 1997), at page 7.

References

Agrawal, A. A., 2003, "Special Feature: Community Genetics: New Insights into Community Ecology by Integrating Population Genetics," *Ecology* 84:543–544.

Agrawal, A. A., E. D. Brodie III, and M. J. Wade , 2001, "On Indirect Genetic Effects in Structured Populations," *American Naturalist* 158:308–323.

Alberts, B., D. Bray, J. Lewis, M. Raff, K. Roberts, and J. D. Watson, 1994, *Molecular Biology of the Cell*. New York: Garland.

Alexander, R., and G. Borgia , 1978, "Group Selection, Altruism, and the Levels of Organization of Life," *Annual Review of Ecological Systematics* 9:449–474.

Allee, W. C., A. E. Emerson, O. Park, T. Park, and K. P. Schmidt, 1949, *Principles of Animal Ecology*. Philadelphia, PA: W. B. Saunders.

Antonovics, J., 2003, "Toward Community Genomics?," *Ecology* 84:598–601.

Aristotle, *Aristotle's Categories and De Interpretatione*. Translated by J. L. Ackrill. Oxford: Oxford University Press, 1963.

Atran, S., 1999, "Itzaj Maya Folkbiological Taxonomy: Cognitive Universals and Cultural Particulars," in D. Medin and S. Atran (editors), *Folkbiology*. Cambridge, MA: MIT Press.

Avital, E., and E. Jablonka, 2000, *Animal Traditions: Behavioural Inheritance in Evolution*. Cambridge: Cambridge University Press.

Baltimore, D., 2001, "Our Genome Unveiled," *Nature* 409:814–816.

Barkow, J., L. Cosmides, and J. Tooby (editors), 1992, *The Adapted Mind: Evolutionary Psychology and Generation of Culture*. New York: Oxford University Press.

Barrett, J. L., and F. C. Keil, 1996, "Conceptualizing a Non-Natural Entity: Anthropomorphism in God Concepts," *Cognitive Psychology* 31:219–247.

Bateson, W., 1906, "The Progress of Genetic Research," delivered as an inaugural address to the Third Conference on Hybridisation and Plant-Breeding. Reprinted in R. C. Punnett (editor), *The Scientific Papers of William Bateson*, Volume II. London: Cambridge University Press, 1928.

Beadle, G. W., and E. L. Tatum, 1941, "Genetic Control of Biochemical Reactions in *Neurospora*," *Proceedings of the National Academy of Science* 21:499–506.

Bechtel, W., and R. Richardson, 1993, *Discovering Complexity: Decomposition and Localization as Strategies in Scientific Research*. Princeton, NJ: Princeton University Press.

Bekoff, M., 1995, "Play Signals as Punctuation: The Structure of Social Play in Canids," *Behaviour* 132:419–429.

Bekoff, M., 1998, "Playing with Play: What Can We Learn About Evolution and Cognition," in D. Cummins and C. Allen (editors), *The Evolution of Mind*. New York: Oxford University Press.

Bering, J. M., 2002, "Intuitive Conceptions of Dead Agents' Minds: The Natural Foundations of Afterlife Beliefs as Phenomenological Boundary," *Journal of Cognition and Culture* 2:263–308.

Bernardes, A. T., and R. M. Zorzenon Dos Santos, 1997, "Immune Network at the Edge of Chaos," *Journal of Theoretical Biology* 186:173–187.

Beurton, P. J., 2000, "A Unified View of the Gene, or How to Overcome Reductionism," in P. J. Beurton, R. Falk, and H-J. Rheinberger (editors), *The Concept of the Gene in Development and Evolution: Historical and Epistemological Perspectives*. New York: Cambridge University Press.

Beurton, P. J., R. Falk, and H-J. Rheinberger (editors), 2000, *The Concept of the Gene in Development and Evolution: Historical and Epistemological Perspectives*. New York: Cambridge University Press.

Black, M., 1962, "Metaphor," in *Models and Metaphors*. Ithaca, NY: Cornell University Press.

Blackburn, S., 1993, *Essays in Quasi-Realism*. New York: Oxford University Press.

Bonner, J. T., 1993, *Life Cycles*. Princeton, NJ: Princeton University Press.

Bourke, A. F. G., and N. R. Franks, 1995, *Social Evolution in Ants*. Princeton, NJ: Princeton University Press.

Boyd, R. N., 1979, "Metaphor and Theory Change: What is 'Metaphor' a Metaphor For?," in A. Ortony (editor), *Metaphor and Thought*, 2nd edition. Cambridge: Cambridge University Press.

Boyd, R. N., 1988, "How to be a Moral Realist," in G. Sayre-McCord (editor), *Essays on Moral Realism*. Ithaca, NY: Cornell University Press.

Boyd, R. N., 1991, "Realism, Anti-foundationalism, and the Enthusiasm for Natural Kinds," *Philosophical Studies* 61:127–148.

Boyd, R. N., 1999, "Homeostasis, Species, and Higher Taxa," in R. A. Wilson (editor), *Species: New Interdisciplinary Essays*. Cambridge, MA: MIT Press.

Boyd, R. N., 1999, "Kinds, Complexity, and Multiple Realization," *Philosophical Studies* 95:67–98.

Brasier, C., 1992, "A Champion Thallus," *Nature* 356:382–383.

Brenner, S.,W. Dover, I. Herskowitz, and R. Thomas, 1990, "Genes and Development: Molecular and Logical Themes," *Genetics* 126:479–486.

Brown, J. H., and B. A. Maurer, 1987, "Evolution of Species Assemblages," *American Naturalist* 130:1–17.

Brown, J. H., and B. A. Maurer, 1989, "Macroecology: The Division of Food and Space Among Species on Continents," *Science* 243:1145–1150.

Buss, L., 1987, *The Evolution of Individuality*. Princeton, NJ: Princeton University Press.

Capra, F., 2002, *The Hidden Connections: Integrating the Biological, Cognitive, and Social Dimensions of Life into a Science of Sustainability*. New York: Doubleday.

Cavender-Bares J., and A. Wilczek, 2003, "Integrating Micro- and Macroevolutionary Processes in Community Ecology," *Ecology* 84:592–597.

Chalupa, L. M., 1995, "The Nature and Nurture of Retinal Ganglion Cell Development," in M. Gazzaniga (editor), *The Cognitive Neurosciences*. Cambridge, MA: MIT Press.

Chase, J. M., and T. M. Knight, 2003, "Community Genetics: Toward a Synthesis," *Ecology* 84:580–582.

Churchland, P. M., 1979, *Scientific Realism and the Plasticity of Mind*. New York: Cambridge University Press.

Clark, A., 1997, *Being There: Putting Brain, Body, and World Together Again*. Cambridge, MA: MIT Press.

Collins, J. P., 2003, "What Can We Learn From Community Genetics?," *Ecology* 84:574–577.

Cracraft, J., 1983, "Species Concepts and Speciational Analysis," *Current Ornithology* 1:159–187. Reprinted in M. Ereshefsky (editor), *The Units of Evolution: Essays on the Nature of Species*. Cambridge, MA: MIT Press.

Craig, D. M., 1982, "Group Selection Versus Individual Selection: An Experimental Analysis," *Evolution* 36:271–282.

Crick, F., 1958, "On Protein Synthesis," in *Symposium of the Society for Experimental Biology* 12:138–163.

Crozier, R. H., and P. Pamilo, 1996, *Evolution of Social Insect Colonies: Sex Allocation and Kin Selection*. New York: Oxford University Press.

Cummins, R., 1983, *The Nature of Psychological Explanation*. Cambridge, MA: MIT Press.

Damuth, J., and I. L. Heisler , 1988, "Alternative Formulations of Multilevel Selection," *Biology and Philosophy* 3:407–430.

Darwin, C., 1964, *On the Origin of Species: A Facsimile of the First Edition*. Cambridge, MA: Harvard University Press. 1st edition, 1859.

Darwin, C., 1981, *The Descent of Man, and Selection in Relation to Sex*. Princeton, NJ: Princeton University Press. 1st edition, 1871.

Dawkins, R., 1982, *The Extended Phenotype: The Gene as the Unit of Selection*. Oxford: Oxford University Press.

Dawkins, R., 1989, "The Evolution of Evolvability," in C. Langton (editor), *Artificial Life*. Sante Fe, NM: Addison-Wesley.

Dawkins, R., 1989, *The Selfish Gene*, 2nd edition. New York: Oxford University Press. 1st edition, 1976.

Dennett, D. C., 1978, "Artificial Intelligence as Philosophy and as Psychology," reprinted in *Brainstorms*. Cambridge, MA: MIT Press, 1978.

Dennett, D. C., 1987, *The Intentional Stance*. Cambridge, MA: MIT Press.

Dennett, D. C., 1991, "Real Patterns," *Journal of Philosophy* 87:27–51.

Dennett, D. C., 1995, *Darwin's Dangerous Idea*. New York: Simon and Schuster.

Dobzhansky, T, 1973, "Nothing in Biology Makes Sense Except in the Light of Evolution," *American Biology Teacher* 35:125–129.

Donald, M., 1991, *The Origins of the Modern Mind*. Cambridge, MA: Harvard University Press.

Donald, M., 2001, *A Mind So Rare: The Evolution of Human Consciousness*. New York: Norton.

Douglas, M., 1986, *How Institutions Think*. Syracuse, NY: Syracuse University Press.

Dresden, M., 1998, "The Klopsteg Memorial Lecture: Fundamentality and Numerical Scales – Diversity and the Structure of Physics," *American Journal of Physics* 66:468–482.

Dugatkin, L. A., 2002, "Will Peace Follow?," *Biology and Philosophy* 17:519–522.

Dugatkin, L. A., and H. K. Reeve, 1994, "Behavioral Ecology and Levels of Selection: Dissolving the Group Selection Controversy," in P. J. B. Slater et al. (editors), *Advances in the Study of Behavior*, Volume 23. New York: Academic Press.

Dupré, J., 1993, *The Disorder of Things: Metaphysical Foundations of the Disunity of Science*. Cambridge, MA: Harvard University Press.

Dupré, J., 1999, "On the Impossibility of a Monistic Account of Species," in R. A. Wilson (editor), *Species: New Interdisciplinary Essays*. Cambridge, MA: MIT Press.

Dupré, J., 2001, *Human Nature and the Limits of Science*. New York: Oxford University Press.

Eldredge, N., 1985, *Unfinished Synthesis*. New York: Oxford University Press.

Eldredge, N., and S. Salthe, 1984, "Hierarchy and Evolution," *Oxford Surveys in Evolutionary Biology* 1:182–206.

Elsasser, W., 1965, *Atom and Organism*. Princeton, NJ: Princeton University Press.

Emerson, A. E., 1939, "Social Coordination and the Superorganism," *American Midland Naturalist* 21:182–209.

Emerson, A. E., 1942, "Basic Comparisons of Human and Insect Societies," *Biological Symposia* 8:163–176.

Emerson, A., 1943, "Ecology, Evolution and Society," *American Naturalist* 77:97–118.

Emerson, A. E., 1946, "The Biological Basis of Social Cooperation," *Illinois Academy of Science Transactions* 29:9–18.

Ereshefsky, M., 1992, (editor), *The Units of Evolution: Essays on the Nature of Species*. Cambridge, MA: MIT Press.

Ereshefsky, M., 1999, "Species and the Linnaean Hierarchy," in R. A. Wilson (editor), *Species: New Interdisciplinary Essays*. Cambridge, MA: MIT Press.

Ereshefsky, M., 2001, *The Poverty of the Linnaean Hierarchy*. New York: Cambridge University Press.

Ewald, P. W., 1994, *The Evolution of Infectious Disease*. New York: Oxford University Press.

Ewald, P. W., 2002, *Plague Time: The New Germ Theory of Disease*. New York: Anchor Books.

Falk, R., 2000, "The Gene – A Concept in Tension," in P. Beurton, R. Falk, and H-J. Rheinberger (editors), *The Concept of the Gene in Development and Evolution: Historical and Epistemological Perspectives*. New York: Cambridge University Press.

Farmer, J. D., and A.d'A. Belin, 1991, abstract for "Artificial Life: The Coming Revolution," in C. Langton et al. (editors), *Artificial Life II, Sante Fe Studies in the Sciences of Complexity 10*. Reading, MA: Addison-Wesley.

Fenner, F., and B. Fantini, 1999, *Biological Control of Vertebrate Pests: The History of Myxomatosis – An Experiment in Evolution*. Oxford: CABI Publishing.

Fenner, F., and F. N. Ratcliffe, 1965, *Myxomatosis*. Cambridge: Cambridge University Press.

Fodor, J. A., 1987, *Psychosemantics: The Problem of Meaning in the Philosophy of Mind.* Cambridge, MA: MIT Press.

Fogle, T., 2000, "The Dissolution of Protein Coding Genes in Molecular Biology," in P. J. Beurton, R. Falk, and H-J. Rheinberger (editors), *The Concept of the Gene in Development and Evolution: Historical and Epistemological Perspectives.* New York: Cambridge University Press.

Frank, S. A., 1997, "The Price Equation, Fisher's Fundamental Theorem, Kin Selection, and Causal Analysis," *Evolution* 51:1712–1729.

Frank, S. A., 1998, *Foundations of Social Evolution.* Princeton, NJ: Princeton University Press.

Futuyma, D. J., 1986, *Evolutionary Biology*, 2nd edition. Sunderland, MA: Sinauer Associates.

Gehring, W. E., 1998, *Master Control Genes in Development and Evolution: The Homeobox Story.* New Haven, CT: Yale University Press.

Ghiselin, M., 1974, "A Radical Solution to the Species Problem," *Systematic Zoology* 23:536–544. Reprinted in M. Ereshefsky (editor), *The Units of Evolution: Essays on the Nature of Species.* Cambridge, MA: MIT Press.

Ghiselin, M., 1997, *Metaphysics and the Origins of Species.* Albany, NY: SUNY Press.

Gibson, J. J., 1979, *The Ecological Approach to Visual Perception.* Boston, MA: Houghton-Mifflin.

Gilbert, S., 1997, *Developmental Biology.* Cambridge, MA: Sinauer Associates, 5th edition.

Gilbert, W., 1978, "Why Genes in Pieces?," *Nature* 271:501.

Godfrey-Smith, P., 1996, *Complexity and the Function of Mind in Nature.* New York: Cambridge University Press.

Godfrey-Smith, P., 1999, "Genes and Codes: Lessons from the Philosophy of Mind?," in V. Hardcastle (editor), *Where Biology Meets Psychology: Philosophical Essays.* Cambridge, MA: MIT Press.

Godfrey-Smith, P., 2000, "Information, Arbitrariness, and Selection: Comments on Maynard Smith," *Philosophy of Science* 67:202–207.

Godfrey-Smith, P., 2000, "The Replicator in Retrospect," *Biology and Philosophy* 15:403–423.

Godfrey-Smith, P., and B. Kerr , 2002, "Group Fitness and Multi-Level Selection: Replies to Commentaries," *Biology and Philosophy* 17: 539–549.

Goodnight, C., 1985, "The Influence of Environmental Variation on Group and Individual Selection in a Cress," *Evolution* 39:545–558.

Goodnight, C., and L. Stevens, 1997, "Experimental Studies of Group Selection: What Do They Tell Us About Group Selection in Nature?," *American Naturalist* 150 (supp.):S59-S79.

Gould, S. J., 2001, "The Evolutionary Definition of Selective Agency, Validation of the Theory of Hierarchical Selection, and Fallacy of the Selfish Gene," in R. S. Singh, C. B. Krimbas, D. B. Paul, and J. Beatty (editors), *Thinking About Evolution: Historical, Philosophical, and Political Perspectives.* New York: Cambridge University Press.

Gould, S. J., 2002, *The Structure of Evolutionary Theory.* Cambridge, MA: Harvard University Press.

Grafen, A., 1984, "Natural Selection, Kin Selection, and Group Selection," in J. Krebs and N. Davies (editors), *Behavioural Ecology: An Evolutionary Approach.* London: Blackwell.

Grantham, T. A., 1995, "Hierarchical Approaches to Macroevolution: Recent Work on Species Selection and the 'Effect Hypothesis,'" *Annual Review of Ecology and Systematics* 26:301–321.

Gray, R. D., 2001, "Selfish Genes or Developmental Systems? Evolution without Interactors and Replicators?," in R. Singh, C. Krimbas, J. Beatty, and D. Paul (editors), *Thinking about Evolution: Historical, Philosophical, and Political Perspectives.* New York: Cambridge University Press.

Griesemer, J., 2000, "Reproduction and the Reduction of Genetics," in P. Beurton, R. Falk, and H-J. Rheinberger (editors), *The Concept of the Gene in Development and Evolution: Historical and Epistemological Perspectives.* New York: Cambridge University Press.

Griffiths, P. E., 1999, "Squaring the Circle? Natural Kinds with Historical Essences," in R. A. Wilson (editor), *Species: New Interdisciplinary Essays.* Cambridge, MA: MIT Press.

Griffiths, P. E., 2001, "Genetic Information: A Metaphor in Search of a Theory," *Philosophy of Science* 68:394–412.

Griffiths, P. E., 2002, "Review of Sober and Wilson 1998," *Mind* 111:178–182.

Griffiths, P. E., in press, "The Fearless Vampire Conservator: Philip Kitcher, Genetic Determinism and the Informational Gene," in C. Rehmann-Sutter and E. M. Neumann-Held (editors), *Genes in Development.* Durham, NC: Duke University Press.

Griffiths, P. E., and R. D. Gray , 1994, "Developmental Systems and Evolutionary Explanation," *Journal of Philosophy* 91: 277–304.

Griffiths, P. E., and R. D. Gray, 2001, "Darwinism and Developmental Systems," in S. Oyama, P. E. Griffiths, and R. D. Gray (editors), *Cycles of Contingency: Developmental Systems and Evolution.* Cambridge, MA: MIT Press.

Haig, D., 1993, "Genetic Conflicts of Human Pregnancy," *Quarterly Review of Biology* 68:495–532.

Hall, B. K., and S. Hörstadius, 1988, *The Neural Crest.* Oxford: Oxford University Press.

Hamilton, W. D., 1963, "The Evolution of Altruistic Behavior," *American Naturalist* 97:354–356.

Hamilton, W. D., 1964, "The Genetical Evolution of Social Behaviour I," *Journal of Theoretical Biology* 7: 1–16.

Hamilton, W. D., 1964, "The Genetical Evolution of Social Behaviour II," *Journal of Theoretical Biology* 7:17–52.

Hamilton, W. D., 1967, "Extraordinary Sex Ratios," *Science* 156:477–488.

Hamilton, W. D., 1996, *Narrow Roads of Gene Land, Volume 1: Evolution of Social Behavior.* New York: Freeman.

Harper, J., 1977, *Population Biology of Plants.* New York: Academic Press.

Heider, F., and M. Simmel , 1944, "An Experimental Study of Apparent Behavior," *American Journal of Psychology* 57:243–259.

Heisler, I. L., and J. Damuth, 1987, "A Method for Analyzing Selection in Hierarchically Structured Populations," *American Naturalist* 130:582–602.

Hesse, M., 1966, *Models and Analogies in Science*. South Bend, IN: University of Notre Dame Press.

Hillis, D., 1991, "Co-evolving Parasites Improve Simulated Evolution as an Optimization Procedure," in C. Langton, C. Taylor, J. D. Farmer, and S. Rasmussen (editors), *Artificial Life II. Santa Fe Institute Studies in the Sciences of Complexity*. Redwood City: CA: Addison-Wesley.

Hrdy, S. B., 1999, *Mother Nature: A History of Mothers, Infants, and Natural Selection*. New York: Pantheon.

Hughes, A., 1979, "A Rose by Any Other Name . . . : On 'Naming of Neurones' by Rowe and Stone," *Brain, Behavior, and Evolution* 16:52–64.

Hull, D., 1965, "The Effect of Essentialism on Taxonomy: Two Thousand Years of Stasis," *British Journal for the Philosophy of Science* 15: 314–326 and 16: 1–18. Reprinted in M. Ereshefsky (editor), *The Units of Evolution: Essays on the Nature of Species*. Cambridge, MA: MIT Press.

Hull, D., 1976, "Are Species Really Individuals?," *Systematic Zoology* 25:174–191.

Hull, D., 1978, "A Matter of Individuality," *Philosophy of Science* 45:335–360. Reprinted in M. Ereshefsky (editor), *The Units of Evolution: Essays on the Nature of Species*. Cambridge, MA: MIT Press.

Hull, D., 1980, "Individuality and Selection," *Annual Review of Ecology and Systematics* 11:311–332.

Hull, D., 1981, "Units of Evolution: A Metaphysical Essay," in U. L. Jensen and R. Harré (editors), *The Philosophy of Evolution*. Brighton, Sussex: Harvester Press.

Hull, D., and M. Ruse (editors), *The Philosophy of Biology*. New York: Oxford University Press.

Hutchins, E., 1995, *Cognition in the Wild*. Cambridge, MA: MIT Press.

Huxley, J., 1912, *The Individual in the Animal Kingdom*. London: Cambridge University Press.

Ingold, T., 2001, "From Complementarity to Obviation: On Dissolving the Boundaries between Social and Biological Anthropology, Archaeology, and Psychology," in S. Oyama, P. E. Griffiths, and R. D. Gray (editors), *Cycles of Contingency: Developmental Systems and Evolution*. Cambridge, MA: MIT Press.

Jablonka, E., 2001, "The Systems of Inheritance," in Susan Oyama, Paul E. Griffiths, and Russell D. Gray (editors), *Cycles of Contingency: Developmental Systems and Evolution*. Cambridge, MA: MIT Press.

Jablonka, E., and M. Lamb, 1995, *Epigenetic Inheritance and Evolution: the Lamarckian Dimension*. New York: Oxford University Press.

Jablonka, E., and M. Lamb , 1998, "Epigenetic Inheritance in Evolution," *Journal of Evolutionary Biology* 11:159–183.

Jablonka, E., and M. Lamb, 2005, *Evolution in Four Dimensions*. Cambridge, MA: MIT Press.

Jablonski, D., 1986, "Larval Ecology and Macroevolution in Marine Invertebrates," *Bulletin of Marine Sciences* 39: 565–587.

Jablonski, D., 1987, "Heritability at the Species Level: Analysis of Geographic Ranges of Cretaceous Mollusks," *Science* 238:360–363.

Jacob, F., 1973, *The Logic of Life*. Princeton, NJ: Princeton University Press.

Jacob, F., and J. Monod , 1961, "On the Regulation of Gene Activity," *Cold Spring Harbor Symposium on Quantitative Biology* 26:193–211.

Jacob, F., and J. Monod, 1961, "Genetic Regulatory Mechanisms in the Synthesis of Proteins," *Journal of Molecular Biology* 3:318–356.

Johannsen, W., 1909, *Elemente der exakten Erblichkeitslehre*. Jena: Gustav Fischer.

Kay, L. E., 2000, *Who Wrote the Book of Life? A History of the Genetic Code*. Palo Alto, CA: Stanford University Press.

Keller, E. F., 1995, *Refiguring Life: Metaphors of Twentieth-Century Biology*. New York: Columbia University Press.

Keller, E. F., 2000, *The Century of the Gene*. Cambridge, MA: Harvard University Press.

Keller, E. F., 2001, "Beyond the Gene but Beneath the Skin," in S. Oyama, P. E. Griffiths, and R. D. Gray (editors), *Cycles of Contingency: Developmental Systems and Evolution*. Cambridge, MA: MIT Press.

Kerr, B., and P. Godfrey-Smith, 2002, "Individualist and Multi-Level Perspectives on Selection in Structured Populations," *Biology and Philosophy* 17:477–517.

Kerr, B., and P. Godfrey-Smith, 2002, "On Price's Equation and Average Fitness," *Biology and Philosophy* 17:551–565.

Kim, J., 1989, "The Myth of Nonreductive Materialism," reprinted in his *Supervenience and Mind*. New York: Cambridge University Press, 1993.

Kim, J., 1992, "Multiple Realization and the Metaphysics of Reduction," *Philosophy and Phenomenological Research* 52:1–26. Reprinted in his *Supervenience and Mind*. New York, Cambridge University Press, 1993.

Kitcher, P., 1984, "Species," *Philosophy of Science* 51:308–333. Reprinted in M. Ereshefsky (editor), *The Units of Evolution: Essays on the Nature of Species*. Cambridge, MA: MIT Press.

Kitcher, P., 1989, "Some Puzzles About Species," in M. Ruse (editor), *What the Philosophy of Biology Is*. Dordrecht, Holland: Kluwer.

Kitcher, P., 1992, "The Naturalists Return," *Philosophical Review* 101:53–114.

Kitcher, P., 1993, "Function and Design," in P. A. French, T. E. Uehling Junior, and H. K. Wettstein (editors), *Midwest Studies of Philosophy, Volume XVIII*. Notre Dame: IN: University of Notre Dame Press.

Kitcher, P., 1996, *The Lives to Come: The Genetical Revolution and Human Possibilities*. New York: Touchstone Books.

Kitcher, P., 1999, "Race, Ethnicity, Biology, Culture," reprinted in his *In Mendel's Mirror*. New York: Oxford University Press, 2003.

Kitcher, P., 2001, "Battling the Undead: How (and How Not) to Resist Genetic Determinism," in R. S. Singh, C. B. Krimbas, D. B. Paul, and J. Beatty (editors), *Thinking About Evolution: Historical, Philosophical, and Political Perspectives*. New York: Cambridge University Press.

Kitcher, P., 2003, *In Mendel's Mirror: Philosophical Reflections on Biology*. New York: Oxford University Press.

Kitcher, P., 2004, "Evolutionary Theory and the Social Uses of Biology," *Biology and Philosophy* 19: 1–15.

Kornblith, H., 1993, *Inductive Inference and its Natural Ground*. Cambridge, MA: MIT Press.

Lack, D., 1954, *The Natural Regulation of Animal Numbers*. New York: Oxford University Press.

Laland, K. N., F. J. Odling-Smee, and M. W. Feldman , 2000, "Niche Construction, Biological Evolution, and Cultural Change," *Behavioral and Brain Sciences* 23: 131–175.

Langton, C., 1986, "Studying Life With Cellular Automata," *Physica D* 22:120–149.

Langton, C., 1989 "Artificial Life," in C. Langton (editor), *Artificial Life: Santa Fe Studies in the Sciences of Complexity* 6. Reading, MA: Addison-Wesley.

Le Bon, G., 1968, *The Crowd: A Study of the Popular Mind*, 2nd edition. Dunwoody, GA: Norman S. Berg. 1st edition, 1895.

Le Douarin, N. M., 1982, *The Neural Crest*. Cambridge: Cambridge University Press.

Le Douarin, N. M., 1987, "The Neural Crest," in G. Adelman (editor), *Encyclopedia of Neuroscience*. Boston, MA: Birkhauser.

Levine, A. J., 1992, *Viruses*. New York: Scientific American Library.

Lewens, T., 2004, *Organisms and Artifacts*. Cambridge, MA: MIT Press.

Lewontin, R., 1970, "The Units of Selection," *Annual Review of Ecology and Systematics* 1:1–18.

Lewontin, R., 1983, "Gene, Organism, and Environment," in D. S. Bendall (editor), *Evolution from Molecules to Men*. Cambridge: Cambridge University Press.

Lewontin, R., 2000, *The Triple Helix: Gene, Organism, and Environment*. Cambridge, MA: Harvard University Press.

Lloyd, E. A., 1994, *The Structure and Confirmation of Evolutionary Theory*. Princeton, NJ: Princeton University Press. Reprint of 1988 edition, New York: Greenwood Press, with a new preface.

Lloyd, E. A., 2001, "Units and Levels of Selection: An Anatomy of the Units of Selection Debates," in R. S. Singh, C. B. Krimbas, D. B. Paul, and J. Beatty (editors), *Thinking About Evolution: Historical, Philosophical, and Political Perspectives*. New York: Cambridge University Press.

Lovelock, J., 1995, *Gaia: A New Look at Life on Earth*. New York: Oxford University Press. Reissue of 1979 edition.

Lycan, W., 1987, *Consciousness*. Cambridge, MA: MIT Press.

Lycan, W. G. (editor), 1998, *Mind and Cognition: A Reader*. New York: Blackwell, 2nd edition. 1st edition, 1990.

Mackie, J. L., 1946, "A Refutation of Morals," *Australasian Journal of Philosophy* 24:77–90.

Mackie, J. L., 1977, *Ethics: Inventing Right and Wrong*. London: Penguin.

Margulis, L., 1993, *Symbiosis in Cell Evolution*. New York: W. H. Freeman.

Marler, P., 1984, "Song Learning: Innate Species Differences in the Learning Process," in P. Marler and H. S. Terrace (editors), *The Biology of Learning*. Berlin: Springer.

Marler, P., 1991, "Differences in Behavioural Development in Closely Related Species: Birdsong," in P. Bateson (editor), *The Development and Integration of Behaviour*. Cambridge: Cambridge University Press.

Matthen, M., and E. Levy, 1984, "Teleology, Error, and the Human Immune System," *Journal of Philosophy* 81:351–372.

Maynard Smith, J., 2000, "The Concept of Information in Biology," *Philosophy of Science* 67:177–194.

Maynard Smith, J., 2002, "Commentary on Kerr and Godfrey-Smith" *Biology and Philosophy* 17:523–527.

Maynard Smith, J., and E. Szathmáry, 1995, *The Major Transitions in Evolution.* New York: Oxford University Press.

Mayr, E., 1970, *Populations, Species, and Evolution.* Cambridge, MA: Harvard University Press.

Mayr, E., 1982, *The Growth of Biological Thought: Diversity, Evolution, and Inheritance.* Cambridge, MA: Harvard University Press.

Mayr, E., and W. B. Provine (editors), 1980, *The Evolutionary Synthesis: Perspectives on the Unification of Biology.* Cambridge, MA: Harvard University Press.

McOuat, G. R., 1996, "Species, Rules, and Meaning: The Politics of Language and the Ends of Definitions in 19th-Century Natural History," *Studies in the History and Philosophy of Science* 27:473–519.

Melander, P., 1993, "How Not to Explain the Errors of the Immune System," *Philosophy of Science* 60:223–241.

Michod, R., 1999, *Darwinian Dynamics: Evolutionary Transitions in Fitness and Individuality.* Princeton, NJ: Princeton University Press.

Michotte, A., 1963, *The Perception of Causality.* Andover, MD: Metheun.

Miki, Y. et al., 1994, "A Strong Candidate for the Breast and Ovarian Cancer Susceptibility Gene BRCA1," *Science* 266:66–73.

Millikan, R. G., 1984, *Language, Thought, and Other Biological Categories.* Cambridge, MA: MIT Press.

Millikan, R. G., 1993, *White Queen Psychology and Other Essays for Alice.* Cambridge, MA: MIT Press.

Millikan, R. G., 2000, *On Clear and Confused Ideas: An Essay About Substance Concepts.* New York: Cambridge University Press.

Mishler, B., and M. Donoghue, 1982, "Species Concepts: The Case for Pluralism," *Systematic Zoology* 31:491–503. Reprinted in M. Ereshefsky (editor), *The Units of Evolution: Essays on the Nature of Species.* Cambridge, MA: MIT Press.

Moore, D. S., 2001, *The Dependent Gene: The Fallacy of 'Nature' vs. 'Nurture.'* New York: Henry Holt.

Moran, N., and P. Baumann, 1994, "Phylogenetics of Cytoplasmically Inherited Microorganisms of Arthropods," *Trends in Ecology and Evolution* 9:15–20.

Morin, P. J., 2003, "Community Ecology and the Genetics of Interacting Species," *Ecology* 84:577–580.

Moss, L., 2003, *What Genes Can't Do.* Cambridge: MIT Press.

Nanney, D. L., and R. A. Wilson, 2001, "Life's Early Years," *Biology and Philosophy* 16:733–746.

Nesse, R. M., and G. C. Williams, 1994, *Why We Get Sick: The New Science of Darwinian Medicine.* New York: Times Books.

Neuhauser, C., D. A. Andow, G. E. Heimpel, G. May, R. G. Shaw, and S. Wagenius, "Community Genetics: Expanding the Synthesis of Ecology and Genetics," *Ecology* 84:545–558.

Neumann-Held, E., 2001, "Let's Talk about Genes: The Process Molecular Gene Concept and Its Context," in S. Oyama, P. E. Griffiths, and R. D. Gray (editors), *Cycles of Contingency: Developmental Systems and Evolution.* Cambridge, MA: MIT Press.

Odling-Smee, F. J., K. N. Laland, and M. W. Feldman, 2003, *Niche Construction: The Neglected Process in Evolution.* Princeton, NJ: Princeton University Press.

Okasha, S., 2001, "Why Won't the Group Selection Controversy Go Away?," *British Journal for the Philosophy of Science* 52:25–50.

Okasha, S., 2002, "Genetic Relatedness and the Evolution of Altruism," *Philosophy of Science* 69:138–149.

Oppenheim, P., and H. Putnam, 1958, "Unity of Science as a Working Hypothesis," in H. Feigl, M. Scriven, and G. Maxwell (editors), *Minnesota Studies in the Philosophy of Science Volume II: Concepts, Theories, and the Mind-Body Problem.* Minneapolis, MN: University of Minnesota Press.

Oyama, S., 2000, *Evolution's Eye: A Systems View of the Biology-Culture Divide.* Durham, NC: Duke University Press.

Oyama, S., 2000, *The Ontogeny of Information: Developmental Systems and Evolution.* 2nd edition. Durham, NC: Duke University Press. 1st edition, 1985.

Oyama, S., P. E. Griffiths, and R. D. Gray (editors), 2001, *Cycles of Contingency: Developmental Systems and Evolution.* Cambridge, MA: MIT Press.

Oyama, S., P. E. Griffiths, and R. D. Gray, "Introduction: What is Developmental Systems Theory?," in their *Cycles of Contingency.* Cambridge, MA: MIT Press.

Parer, I., 1995, "Relationship Between Survival Rate and Survival Time of Rabbits, *Oryctolagus cuniculus* (L.), Challenged with Myxoma Virus," *Australian Journal of Zoology* 43:303–311.

Parer, I.,W. R. Sobey, D. Conolly, and R. Morton , 1995, "Sire Transmission of Acquired Resistance to Myxomatosis," *Australian Journal of Zoology* 43: 459–465.

Park, T., 1932, "Studies in Population Physiology: The Relation of Numbers to Initial Population Growth in the Flour Beetle *Tribolium Confusum* Duval," *Ecology* 13:172–181

Paterson, H. H., 1985, "The Recognition Concept of Species," in E. S. Vrba (editor), *Species and Speciation.* Pretoria: Transvaal Museum Monograph No.4, 195. Reprinted in M. Ereshefsky (editor), *The Units of Evolution: Essays on the Nature of Species.* Cambridge, MA: MIT Press.

Portin, P., 1993, "The Concept of the Gene: Short History and Present Status," *Quarterly Review of Biology* 68:173–223.

Putnam, H., 1962, "The 'Analytic' and the 'Synthetic,' " reprinted in *Mind, Language, and Reality: Philosophical Papers, Volume 2.* Cambridge: Cambridge University Press, 1975.

Ray, T., 1991, "An Approach to the Synthesis of Life," in C. Langton, C. Taylor, J. D. Farmer, and S. Rasmussen (editors), *Artificial Life II. Santa Fe Institute Studies in the Sciences of Complexity.* Redwood City: CA: Addison-Wesley.

Reeve, H. K., 2000, "Multi-Level Selection and Human Cooperation," *Evolution and Human Behavior* 21:65–72.

Reeve, H. K., and L. Keller, 1999, "Levels of Selection: Burying the Units-of-Selection Debate and Unearthing the Crucial New Issues," in L. Keller (editor), *Levels of Selection in Evolution.* Princeton, NJ: Princeton University Press.

Rheinberger, H-J., 2000, "Gene Concepts: Fragments from the Perspective of Molecular Biology," P. J. Beurton, R. Falk, and H-J. Rheinberger (editors), *The Concept of the Gene in Development and Evolution: Historical and Epistemological Perspectives.* New York: Cambridge University Press.

Ricklefs, R. E., 2003, "Genetics, Evolution, and Ecological Communities," *Ecology* 84:588–591.

Ridley, M., 2001, *Mendel's Demon: Gene Justice and the Complexity of Life*. London: Phoenix.

Ritter, W., 1919, *The Unity of the Organism*. 2 volumes. Boston, MA: R. G. Badger.

Robert, J., 2001, "Interpreting the Homeobox: Metaphors of Gene Action and Activation in Development and Evolution," *Evolution and Development* 3:287–295.

Robert, J., B. Hall, and W. M. Olson, 2001, "Bridging the Gap Between Developmental Systems Theory and Evolutionary Developmental Biology," *BioEssays* 23:954–962.

Rosenberg, A., 1992, "Altruism: Theoretical Contexts," in E. F. Keller and E. A. Lloyd (editors), *Keywords in Evolutionary Biology*. Cambridge, MA: Harvard University Press.

Rosencrantz, G. S., 2001, "What is Life?," in T. Y. Cao (editor), *Proceedings of the World Congress of Philosophy*, Volume 10. Bowling Green, OH: Philosophy Documentation Center.

Rowe, M. H., and J. Stone, 1977: "Naming of Neurones: Classification and Naming of Cat Retinal Ganglion Cells," *Brain, Behavior, and Evolution* 14:185–216.

Rowe, M. H., and J. Stone, 1979, "The Importance of Knowing One's Own Presuppositions," *Brain, Behavior, and Evolution* 16:65–80.

Rowe, M. H., and J. Stone , 1980, "The Interpretation of Variation in the Classification of Nerve Cells," *Brain, Behavior, and Evolution* 17:123–151.

Rowe, M. H., and J. Stone , 1980, "Parametric and Feature Extraction Analyses of the Receptive Fields of Visual Neurones: Two Streams of Thought in the Study of a Sensory Pathway," *Brain, Behavior, and Evolution* 17:103–122.

Runeson, S., 1977, "On the Possibility of 'Smart' Perceptual Mechanisms," *Scandinavian Journal of Psychology* 18:172–179.

Sarkar, S., 2000, "Information in Genetics and Developmental Biology: Comments on Maynard Smith," *Philosophy of Science* 67:208–213.

Schmid-Hempel, P., 1998, *Parasites in Social Insects*. Princeton, NJ: Princeton University Press.

Schopf, J. W., 1999, *Cradle of Life: The Discovery of Earth's Earliest Fossils*. Princeton, NJ: Princeton University Press.

Schrödinger, E., 1944, *What is Life?* Cambridge: Cambridge University Press.

Seeley, D., 1995, *The Wisdom of the Hive: The Social Physiology of Honey Bee Colonies*. Cambridge, MA: Harvard University Press.

Shapiro, L., 2000, "Multiple Realizations," *Journal of Philosophy* 97:635–654.

Shapiro, L., 2004, *The Mind Incarnate*. Cambridge, MA: MIT Press.

Singh, R. S., C. B. Krimbas, D. B. Paul, and J. Beatty (editors), 2001, *Thinking about Evolution: Historical, Philosophical, and Political Perspectives*. New York: Cambridge University Press.

Smith, M. L., J. N. Bruhn, and J. B. Anderson, 1992, "The Fungus *Armillaria bulbosa* is Among the Largest and Oldest Living Organisms," *Nature* 356:428–433.

Smith, R., 1997, *The Norton History of the Human Sciences*. New York: Norton.

Smocovitis, V. B., 1996, *Unifying Biology: The Evolutionary Synthesis and Evolutionary Biology*. Princeton, NJ: Princeton University Press.

Sober, E., 1980, "Evolution, Population Thinking, and Essentialism," *Philosophy of Science* 47:350–383. Reprinted in M. Ereshefsky (editor), *The Units of Evolution: Essays on the Nature of Species*. Cambridge, MA: MIT Press.

Sober, E., 1984, *The Nature of Selection.* Cambridge, MA: MIT Press.

Sober, E., 1993, *Philosophy of Biology.* Boulder, CO: Westview Press.

Sober, E., and R. Lewontin, 1982, "Artifact, Cause, and Genic Selection," *Philosophy of Science* 49:157–180.

Sober, E., and D. S. Wilson, 1994, "A Critical Review of Philosophical Work on the Units of Selection Problem," *Philosophy of Science* 61:534–555.

Sober, E., and D. S. Wilson, 1998, *Unto Others: The Evolution and Psychology of Unselfish Behavior.* Cambridge, MA: Harvard University Press.

Sober, E., and D. S. Wilson, 2002, "Perspectives and Parameterizations: Commentary on Benjamin Kerr and Peter Godfrey-Smith's 'Individualist and Multi-Level Perspectives on Selection in Structured Populations,'" *Biology and Philosophy* 17:529–537.

Sobey, W. R., and D. Conolly , 1986, "Myxomatosis: Non-genetic Aspects of Resistance to Myxomatosis in Rabbits, *Oryctolagus cuniculus,*" *Australian Wildlife Research* 13:177–187.

Sokal, R. R., and T. J. Crovello, 1970, "The Biological Species Concept: A Critical Evaluation," *American Naturalist* 104:12–153. Reprinted in M. Ereshefsky (editor), *The Units of Evolution: Essays on the Nature of Species.* Cambridge, MA: MIT Press.

Sokal, R. R., and P. H. A. Sneath, 1963, *Principles of Numerical Taxonomy.* San Francisco: Freeman.

Sonneborne, T., 1957, "Breeding Systems, Reproductive Methods, and Species Problems in Protozoa," in E. Mayr (editor), *The Species Problem.* Washington, DC: American Association for the Advancement of Science, Publication 50.

Spencer, H. 1852, "A Theory of Population, Deduced from the General Law of Animal Fertility," *Westminster Review* 1:468–501.

Spencer, H., 1866, *The Principles of Psychology.* Volume I. New York: D. Appleton.

Spencer, H., 1898, *The Principles of Biology.* Volume I. New York: D. Appleton. Revised and enlarged edition. 1st edition, 1866.

Sperber, D., D. Premack, and A. Premack (editors), 1995, *Causal Cognition: A Multidisciplinary Debate.* New York: Oxford University Press.

Stanley, S. M., 1979, *Macroevolution: Pattern and Process.* San Francisco: W. H. Freeman.

Stanley, S. M., 1981, *The New Evolutionary Timetable: Fossils, Genes, and the Origin of Species.* New York: Basic Books.

Sterelny, K., 1996, "The Return of the Group," *Philosophy of Science* 63: 562–584.

Sterelny, K., 2000, "The 'Genetic Program' Program: A Commentary on Maynard Smith on Information in Biology," *Philosophy of Science* 67:195–201.

Sterelny, K., 2003, "Last Will and Testament: Stephen Jay Gould's *The Structure of Evolutionary Theory,*" *Philosophy of Science* 70:255–263.

Sterelny, K., and P. Kitcher, 1988, "The Return of the Gene," *Journal of Philosophy* 85:339–361.

Svitil, K., 1993, "Fungus Among Us," *Discover* 69:69–70.

Tauber, A., 1994, *The Immune Self: Theory or Metaphor?* New York: Cambridge University Press.

Templeton, A., 1989, "The Meaning of Species and Speciation: A Genetic Perspective," in D. Otte and J. A. Endler (editors), *Speciation and its Consequences.*

Sunderland, MA: Sinauer Associates. Reprinted in M. Ereshefsky (editor), *The Units of Evolution: Essays on the Nature of Species.* Cambridge, MA: MIT Press.

Thompson, P., 2001, "'Organization,' 'Population,' and Mayr's Rejection of Essentialism in Biology," in D. Sfendoi-Mentzou, J. Hattiangadi, and D. M. Johnson (editors), *Aristotle and Contemporary Science, Volume 2.* New York: Peter Lang.

Trivers, R., 1971, "The Evolution of Reciprocal Altruism," *Quarterly Review of Biology* 46:35–57.

Trivers, R., 1974, "Parent-Offspring Conflict," *American Zoologist* 14:249–274.

Trout, J. D., 1988, "Attribution, Content, and Method: A Scientific Defense of Commonsense Psychology." Ph.D. thesis, Cornell University.

Turner, J. S., 2000, *The Extended Organism: The Physiology of Animal-Built Structures.* Cambridge, MA: Harvard University Press.

Tyner, C. F., 1975, "The Naming of Neurons: Applications of Taxonomic Theory to the Study of Cellular Populations," *Brain, Behavior, and Evolution* 12:75–96.

van Inwagen, P., 1990, *Material Beings.* Ithaca, NY: Cornell University Press.

von Neumann, J., and O. Morgenstern, 1953 [1944], *The Theory of Games and Economic Behaviour.* Princeton, NJ: Princeton University Press, 2nd edition.

Vrba, E. S., 1984, "What is Species Selection?" *Systematic Zoology* 33:318–328.

Vrba, E. S., 1989, "Levels of Selection and Sorting with Special Reference to the Species Level," in P. H. Harvey and L. Partridge (editors), *Oxford Surveys in Evolutionary Biology, Volume 6.* Oxford: Oxford University Press.

Vrba, E. S., and S. J. Gould, 1986, "The Hierarchical Expansion of Sorting and Selection: Sorting and Selection Cannot Be Equated," *Paleobiology* 12: 217–228.

Wade, M. J., 1977, "An Experimental Study of Group Selection," *Evolution* 31: 134–153.

Wade, M. J., 1978, "A Critical Review of the Models of Group Selection," *Quarterly Review of Biology* 53:101–114.

Wade, M. J., 1980, "An Experimental Study of Kin Selection," *Evolution* 34: 844–855.

Wade, M. J., 1982, "Group Selection: Migration and the Differentiation of Small Populations," *Evolution* 36:944–961.

Wade, M. J., 2001, "Infectious Speciation," *Nature* 409:675–677.

Wade, M. J., 2003, "Community Genetics and Species Interactions," *Ecology* 84:583–585.

Walsh, D. M., 1999, "Alternative Individualism," *Philosophy of Science* 66:628–648.

Wheeler, Q. D., and R. Meier (editors), 2000, *Species Concepts and Phylogenetic Theory: A Debate.* New York: Columbia University Press.

Whitham, T. G., et al., 2003, "Community and Ecosystem Genetics: A Consequence of the Extended Phenotype," *Ecology* 84:559–573.

Wigglesworth, V. B., 1964, *The Life of Insects.* London: Wiedenfeld and Nicolson.

Wiley, E. O., 1978, "The Evolutionary Species Concept Reconsidered," *Systematic Zoology* 27:17–26. Reprinted in M. Ereshefsky (editor), *The Units of Evolution: Essays on the Nature of Species.* Cambridge, MA: MIT Press.

Williams, C. K., and R. J. Moore, 1991, "Inheritance of Acquired Immunity to Myxomatosis," *Australian Journal of Zoology* 39:307–311.

Williams, G. C., 1966, *Adaptation and Natural Selection: A Critique of some Current Evolutionary Thought.* Princeton, NJ: Princeton, NJ.

Williams, G. C., 1985, "A Defense of Reductionism in Evolutionary Biology," in R. Dawkins and M. Ridley (editors), *Oxford Surveys in Evolutionary Biology*, Volume 2. Oxford: Oxford University Press.

Williams, G. C., 1992, *Natural Selection: Domains, Levels, and Challenges*. New York: Oxford University Press.

Wilson, D. S., 1975, "A Theory of Group Selection," *Proceedings of the National Academy of Sciences USA* 72:143–146.

Wilson, D. S., 1977, "Structured Demes and the Evolution of Group Advantageous Traits," *American Naturalist* 111:157–185.

Wilson, D. S., 1980, *The Natural Selection of Populations and Communities*. Menlo Park, CA: Benjamin Cummins.

Wilson, D. S., 1983, "The Group Selection Controversy: History and Current Status," *Annual Review of Ecology and Systematics* 14:159–187.

Wilson, D. S., 1988, "Holism and Reductionism in Evolutionary Ecology," *Oikos* 53:269–273.

Wilson, D. S., 1989, "Levels of Selection: An Alternative to Individualism in the Human Sciences," *Social Networks* 11:257–272. Reprinted in E. Sober (editor), *Conceptual Issues in Evolutionary Biology*. Cambridge, MA: MIT Press, 2nd edition, 1993.

Wilson, D. S., 1997, "Altruism and Organism: Disentangling the Themes of Multilevel Selection Theory," *American Naturalist* 150 (supp.):S122-S134.

Wilson, D. S., 1997, "Incorporating Group Selection into the Adaptationist Program: A Case Study Involving Human Decision Making," in J. Simpson and D. Kendrick (editors), *Evolutionary Social Psychology*. Hillsdale, NJ: Erlbaum.

Wilson, D. S., 1999, "A Critique of R. D. Alexander's Views on Group Selection," *Biology and Philosophy* 14:431–449.

Wilson, D. S., 2002, *Darwin's Cathedral: Evolution, Religion, and the Nature of Society*. Chicago: University of Chicago Press.

Wilson, D. S., and L. Dugatkin, 1992, "Altruism: Contemporary Debates," E. F. Keller and E. A. Lloyd (editors), *Keywords in Evolutionary Biology*. Cambridge, MA: Harvard University Press.

Wilson, D. S., and E. Sober, 1989, "Reviving the Superorganism," *Journal of Theoretical Biology* 136:337–356.

Wilson, D. S., and E. Sober, 1994, "Reintroducing Group Selection to the Human Behavioral Sciences," *Behavioral and Brain Sciences* 17:585–654.

Wilson, D. S., and W. Swenson, 2003, "Community Genetics and Community Selection," *Ecology* 84:586–588.

Wilson, D. S., C. Wilczynski, A. Wells, and L. Weiser, 2000, "Gossip and Other Aspects of Language as Group-Level Adaptations," in C. Heyes and L. Huber (editors), *Evolution and Cognition*. Cambridge, MA: MIT Press.

Wilson, E. B., 1900, *The Cell in Development and Inheritance*. London, Macmillan.

Wilson, E. O., 1971, *The Insect Societies*. Cambridge, MA: Belknap Press.

Wilson, E. O., 2000, *Sociobiology: The New Synthesis*. Cambridge, MA: Harvard University Press, 2nd edition. 1st edition, 1975.

Wilson, J., 1999, *Biological Individuality: The Identity and Persistence of Living Entities*. New York: Cambridge University Press.

Wilson, R. A., 1994, "Wide Computationalism," *Mind* 103:351–372.

Wilson, R. A., 1995, *Cartesian Psychology and Physical Minds: Individualism and the Sciences of the Mind.* New York: Cambridge University Press.

Wilson, R. A., 1996, "Promiscuous Realism," *British Journal for the Philosophy of Science* 47:303–316.

Wilson, R. A., 1999, "The Individual in Biology and Psychology," in V. Hardcastle (editor), *Where Biology Meets Psychology: Philosophical Essays.* Cambridge, MA: MIT Press.

Wilson, R. A., 1999, "Realism, Essence, and Kind," in R. A. Wilson (editor), *Species: New Interdisciplinary Essays.* Cambridge, MA: MIT Press.

Wilson, R. A., 2000, "The Mind Beyond Itself," in D. Sperber (editor), *Metarepresentations: A Multidisciplinary Perspective.* New York: Oxford University Press.

Wilson, R. A., 2004, *Boundaries of the Mind: The Individual in the Fragile Sciences: Cognition.* New York: Cambridge University Press.

Winsor, M. P., 2003, "Non-essentialist Methods in Pre-Darwinian Taxonomy," *Biology and Philosophy* 18:387–400.

Wittgenstein, L., 1953, *Philosophical Investigations.* Oxford: Blackwell.

Woese, C. R., 1998, "The Universal Ancestor," *Proceedings of the National Academy of Sciences USA* 95:6854–6859.

Woese, C. R., 2000, "Interpreting the Universal Phylogenetic Tree," *Proceedings of the National Academy of Sciences USA* 97:8392–8396.

Woese, C. R., O. Kandler, and M. L. Wheeler, 1990, "Towards a Natural System of Organisms: Proposal for the Domains Archaea, Bacteria, and Eucarya," *Proceedings of the National Academy of Sciences USA* 87:4576–4579.

Wolf, J. B., E. D. Brodie III, and M. J. Wade, 2004, "Evolution When the Environment Contains Genes," in T. J. DeWitt and S. M. Scheiner (editors), *Phenotypic Plasticity: Functional and Conceptual Approaches.* New York: Oxford University Press.

Wynne-Edwards, V. C., 1962, *Animal Dispersion in Relation to Social Behaviour.* Edinburgh: Oliver and Boyd.

Index

Printed in the United States
by Baker & Taylor Publisher Services